TM 9-285

US ARMY TECHNICAL MANUAL

SHOTGUNS, ALL TYPES

1942

WORLD WAR II
CIVILIAN REFERENCE EDITION

UNABRIDGED FIELD MANUAL ON VINTAGE AND CLASSIC SHOTGUNS
FOR HUNTING, TRAP, SKEET, AND DEFENSE FROM THE WARTIME ERA

U.S. WAR DEPARTMENT

Doublebit Press

New content, introduction, cover design, and annotations
Copyright © 2020 by Doublebit Press. All rights reserved.

Doublebit Press is an imprint of Eagle Nest Press
www.doublebitpress.com
Cherry, IL, USA

Original content under the public domain; unrestricted for civilian distribution. Originally published in 1942 US War Department.

This title, along with other Doublebit Press books are available at a volume discount for youth groups, clubs, or reading groups. Contact Doublebit Press at info@doublebitpress.com for more information.

Military Outdoors Skills Series: Volume 6

Doublebit Press Civilian Reference Edition ISBNs
Hardcover: 978-1-64389-154-5
Paperback: 978-1-64389-155-2

Doublebit Press, or its employees, authors, and other affiliates, assume no liability for any actions performed by readers or any damages that might be related to information contained in this book. Some of the material in this book may be outdated by modern standards. This text has been published for historical study and for personal literary enrichment. Remember to be safe with any activity that you do in the outdoors and to help do your part to preserve and be a good steward of our great American wild lands.

The Military Outdoors Skills Series
Historic Field Manuals and Military Guides
on Outdoors Skills and Travel

Military manuals contain essential knowledge about outdoors life, thriving while in the field, and self-sufficiency. Unfortunately, many great military books, field manuals, and technical guides over the years have become less available and harder to find. These have either been rescinded by the armed forces or are otherwise out of print due to their age. This does not mean that these manuals are worthless or "out of date" – in fact, the opposite is true! It is true that the US Military frequently updates its manuals as its protocols frequently change based on the current times and combat situations that our armed services face. However, the knowledge about the outdoors over the entire history of military publications is timeless!

By publishing the **Military Outdoors Skills Series**, it is our goal at Doublebit Press to do what we can to preserve and share valuable military works that hold timeless knowledge about outdoors life, navigation, and survival. These books include official unrestricted texts such as army field manuals (the FM series), technical manuals (the TM series), and other military books from the Air Force, Navy, and texts from before 1900. Through remastered reprint editions of military handbooks and field manuals, outdoors enthusiasts, bushcrafters, hunters, scouts, campers, survivalists, nature lore experts, and military historians can preserve the time-tested skills and institutional knowledge that was learned through hard lessons and training by the U.S. Military and our expert soldiers.

Soldiers were the original campers and survivalists! Because of this, military field manuals about outdoors life contain essential knowledge about thriving in the wilds. This book is not just for soldiers!

This book is an important contribution to outdoors literature and has important historical and collector value toward preserving the American outdoors tradition. The knowledge it holds is an invaluable reference for practicing skills related to thriving in the outdoors. Its chapters thoroughly discuss some of the essential building blocks of outdoors knowledge that are fundamental but may have been forgotten as equipment gets fancier and technology gets smarter. In short, this book was chosen for Historic Edition printing because much of the basic skills and knowledge it contains could be forgotten or put to the wayside in trade for more modern conveniences and methods.

Although the editors at Doublebit Press are thrilled to have comfortable experiences in the woods and love our high-tech and light-weight equipment, we are also realizing that the basic skills taught by the old experts are more essential than ever as our culture becomes more and more hooked on digital technology. We don't want to risk forgetting the important steps, skills, or building blocks involved with thriving in the outdoors. This Civilian Reference Edition reprint represents a collection of military handbooks and field manuals that are essential contributions to the American outdoors tradition despite originating with the military. In the most basic sense, these books are the collection of experiences by the great experts of outdoors life: our countless expert soldiers who learned to thrive in the backwoods, deserts, extreme cold environments, and jungles of the world.

With technology playing a major role in everyday life, sometimes we need to take a step back in time to find those basic building blocks used for gaining mastery – the things that we have luckily not completely lost and has been recorded in books over the last two centuries. These skills aren't forgotten, they've just been shelved. *It's time to unshelve them once again and reclaim the lost knowledge of self-sufficiency.*

Based on this commitment to preserving our outdoors heritage, we have taken great pride in publishing this book as a complete original work. We hope it is worthy of both study and collection by outdoors folk in the modern era of outdoors and traditional skills life.

Unlike many other photocopy reproductions of classic books that are common on the market, this Historic Edition does not simply place poor photography of old texts on our pages and use error-prone optical scanning or computer-generated text. We want our work to speak for itself, and reflect the quality demanded by our customers who spend their hard-earned money. With this in mind, each Historic Edition book that has been chosen for publication is carefully remastered from original print books, *with the Doublebit Civilian Reference Edition printed and laid out in the exact way that it was presented at its original publication.* We provide a beautiful, memorable experience that is as true to the original text as best as possible, but with the aid of modern technology to make as beautiful a reading experience as possible for books that are typically over a century old. Military historians and outdoors enthusiasts alike are sure to appreciate the care to preserve this work!

Because of its age and because it is presented in its original form, the book may contain misspellings, inking errors, and other print blemishes that were common for the age. However, these are exactly the things that we feel give the book its character, which we preserved in this Historic Edition. During digitization, we ensured that each illustration in the text was clean and sharp with the least amount of loss from being copied and digitized as possible. Full-page plate illustrations are presented as they were found, often including the extra blank page that was often behind a plate. For the covers, we use the original cover design to give the book its original feel. We are sure you'll appreciate the fine touches and attention to detail that your Historic Edition has to offer.

For outdoors and military history enthusiasts who demand the best from their equipment, the Doublebit Press Civilian Reference Edition reprint of this military manual was made with you in mind. Both important and minor details have equally both been accounted for by our publishing staff, down to the cover, font, layout, and images. It is the goal of Doublebit Civilian Reference Edition series to preserve outdoors heritage, but also be cherished as collectible pieces, worthy of collection in any outdoorsperson's library and that can be passed to future generations.

that has been chosen for publication is carefully remastered from original print books, *with the Doublebit Civilian Reference Edition printed and laid out in the exact way that it was presented at its original publication.* We provide a beautiful, memorable experience that is as true to the original text as best as possible, but with the aid of modern technology to make as beautiful a reading experience as possible for books that are typically over a century old. Military historians and outdoors enthusiasts alike are sure to appreciate the care to preserve this work!

Because of its age and because it is presented in its original form, the book may contain misspellings, inking errors, and other print blemishes that were common for the age. However, these are exactly the things that we feel give the book its character, which we preserved in this Historic Edition. During digitization, we ensured that each illustration in the text was clean and sharp with the least amount of loss from being copied and digitized as possible. Full-page plate illustrations are presented as they were found, often including the extra blank page that was often behind a plate. For the covers, we use the original cover design to give the book its original feel. We are sure you'll appreciate the fine touches and attention to detail that your Historic Edition has to offer.

For outdoors and military history enthusiasts who demand the best from their equipment, the Doublebit Press Civilian Reference Edition reprint of this military manual was made with you in mind. Both important and minor details have equally both been accounted for by our publishing staff, down to the cover, font, layout, and images. It is the goal of Doublebit Civilian Reference Edition series to preserve outdoors heritage, but also be cherished as collectible pieces, worthy of collection in any outdoorsperson's library and that can be passed to future generations.

TM 9-285

TECHNICAL MANUAL }
TM 9-285

WAR DEPARTMENT
Washington, September 21, 1942

SHOTGUNS, ALL TYPES

Prepared under the direction of the
Chief of Ordnance

CONTENTS

			Paragraphs	Pages
Section	I.	Introduction	1– 3	2– 9
	II.	Winchester Shotgun, 12-Gage, M97	4– 11	10– 37
	III.	Winchester Shotgun, 12-Gage, M12	12– 19	38– 58
	IV.	Stevens Shotgun, 12-Gage, M620A, M520, and M620	20– 27	59– 87
	V.	Ithaca Shotgun, 12-Gage, M37	28– 35	88–111
	VI.	Remington Shotgun, 12-Gage, M10	36– 43	112–136
	VII.	Remington Shotgun, 12-Gage, M31	44– 51	137–160
	VIII.	Remington Shotgun, 12-Gage, M11 and Sportsman	52– 60	161–192
	IX.	Savage Shotgun, 12-Gage, M720	61– 69	193–223
	X.	Stoppages and immediate action	70– 74	224–231
	XI.	Special maintenance	75– 78	232–235
	XII.	Materiel affected by gas	79– 82	236–238
	XIII.	Ammunition	83– 98	239–248
	XIV.	References	99–100	249–250
Index				251–257

TM 9-285
1-2

SHOTGUNS, ALL TYPES

Section I

INTRODUCTION

	Paragraph
Scope	1
Arrangement of manual	2
General	3

1. SCOPE.

a. This technical manual is intended to serve temporarily (pending the publication of a more complete revision) to give information and guidance to personnel of the using arms charged with the operation, maintenance and minor repair of this materiel.

b. This manual contains in brief the available information necessary for the identification, operation, care, and cleaning of the shotguns listed below. In addition is included the disassembly and assembly of the guns for the purpose of cleaning and lubrication, and available information on ammunition.

Shotguns covered in this bulletin are as follows:

Winchester Repeater, 12-Gage, M97	Remington Repeater, 12-Gage, M10
Winchester Repeater, 12-Gage, M12	Remington Repeater, 12-Gage, M31
Stevens Repeater, 12-Gage, M620A	Remington Auto-loading, 12-Gage, M11
Stevens Repeater, 12-Gage, M520	Remington Auto-loading, 12-Gage, Sportsman
Stevens Repeater, 12-Gage, M620	Savage, Auto-loading, 12-Gage, M720
Ithaca Repeater, 12-Gage, M37	

c. Disassembly, assembly, and such repairs as may be handled by using arm personnel will be undertaken only under the supervision of an officer or the chief mechanic.

d. In all cases where the nature of the repair, modification, or adjustment is beyond the scope or facilities of the unit, the responsible ordnance service should be informed in order that trained personnel with suitable tools and equipment may be provided, or proper instructions issued.

2. ARRANGEMENT OF MANUAL.

a. The shotguns covered in this manual are of various makes, models, and types. In some cases different models of the same make differ widely

INTRODUCTION

in design, in others the differences are principally in detail of design. Each make of gun herein is treated separately and where there is extreme variation in model, the model is treated as a separate gun. Where slight variations occur in different models of the same make, they are grouped and the differences explained as they occur.

b. Instructions for disassembly and assembly, and special care and maintenance are covered in the section pertaining to the gun and in section XI; while cleaning, lubrication, and general maintenance, which are more or less common to all the guns, are covered in section II, covering the Winchester M97 gun, and can be applied in general to the other guns as indicated.

c. A general description of the gun for identification, together with such identification marks as may be found upon the gun, are given at the beginning of the section pertaining to the gun in question.

3. GENERAL.

a. The repeating shotguns covered in this manual are of two general types, the slide action, sometimes termed pump action, and the autoloading or semiautomatic. The autoloading gun is often called an automatic, which is incorrect as the trigger must be pulled for each shot.

b. As already explained, different models of the same make of gun may vary in design in whole or in part. Also guns of the same make and model but of different grades may vary slightly in design. In addition some guns of the same make, model, and grade but of various dates of manufacture may have slight variations in design. Such variations are dealt with as far as possible herein. Other variations which may appear must be dealt with as such.

c. Due to absence of standard ordnance nomenclature for the guns covered in this technical manual, with the exception of the Ithaca M37, the parts and assemblies are given the nomenclature supplied by the manufacturer and appearing in their parts lists. Therefore parts and assemblies of, for example, a Winchester gun, may be referred to by a different name than similar parts of a Remington gun. For example, the slide handle, operating handle, and fore end refer to similar parts on different makes of guns. The Ithaca Gun M37 has been given standard nomenclature by the Ordnance Department and this nomenclature is used herein, and will differ in some respects from that appearing in the manufacturer's parts list.

d. The word "shell" is standard nomenclature for the shotgun cartridge. The word "shell" has therefore been used throughout this technical manual and substituted for the word "cartridge" appearing in the manufacturer's parts lists. Therefore when identifying parts referred to herein, this fact should be borne in mind.

SHOTGUNS, ALL TYPES

e. The word "choke" refers to the boring of the barrel, which varies in degree from full cylinder to full choke. In this technical manual three degrees of boring only are referred to, full cylinder (usually referred to as cylinder), improved cylinder, and full choke. The bore of the shotgun barrel has two diameters, the chamber diameter, and the true bore diameter. The chamber diameter is greater than the true bore diameter, and these two diameters are joined by a tapered section usually termed the forcing cone. In a full cylinder gun the true bore diameter extends from the forward end of the forcing cone to the end of the muzzle. Choking is usually accomplished by boring the barrel so that the diameter of the bore near the muzzle end is slightly less than that of the true bore. Diameter of true bore of a 12-gage shotgun is 0.729 inch. The degree of choke in a barrel is measured by the dispersion of the pellets contained in the shot charge at a given distance from the muzzle. This dispersion is measured in the percentage of the number of shot pellets contained in the charge, which will be contained within a 30-inch circle at 40 yards distance from the muzzle. For a cylinder barrel, this will be 40 percent, for an improved cylinder barrel, 50 percent, and for a full choke barrel, 75 percent.

f. The term "hammerless" as applied to the guns in this manual refers to the type of firing mechanism. The Winchester Gun M97 is termed a hammer gun due to the fact that the hammer is visible and operative outside the receiver. The other guns covered in this manual are termed hammerless as the hammers are wholly enclosed within the receiver and are thus not manually operative.

g. The nomenclature of the slide handle, by which the slide action guns in this technical manual are operated, differs for each gun. In the Winchester gun it is called the action slide handle while in the Remington gun it is called the fore end, and in the Ithaca the slide handle, etc. This assembly, however, is basically the same for all the guns, and is composed of three main parts: the slide handle tube which slides on the magazine tube and on which the wooden slide handle is assembled, and the slide handle bar which extends from the rear of the tube into the receiver and connects the handle with the slide or operating mechanism. The tube and bar are integral, welded, or riveted together, according to the make of gun.

h. The term "clockwise" and "counterclockwise" is used in connection with the turning of screwed-in parts to denote the direction of turn. The diameter of the part is considered as the face of a clock and the turn when clockwise is in the direction the hands of the clock would normally travel. Counterclockwise is the opposite.

i. The term "take-down" applies to guns so constructed that the barrel, or barrel magazine and action slide handle group can easily be removed

INTRODUCTION

from the receiver without the use of tools. This construction facilitates cleaning and transportation. The term "solid frame," as used in this manual, refers to guns which either through basic design or the assembly of the bayonet attachment to the barrel are not easily taken down without tools. In the case of the Winchester M97 solid-frame design gun of early manufacture, the barrel and magazine are screwed directly into the receiver and are not to be removed except for repair. In the later design of this gun (par. 4 c) the barrel and magazine are of the take-down type but are locked into the receiver at manufacture and should not be removed. In these Winchester guns, however, it is necessary to remove the bayonet attachment and magazine in order to remove the slide handle, which must be removed in order to remove the groups from the receiver for cleaning when necessary. Therefore, the removal and replacement of these parts is explained. The Winchester M12 (solid-frame gun) when issued, will be assembled similarly to the Winchester M97 (solid-frame gun of later manufacture) (par. 12 b). The other guns covered in this manual are basically of the take-down design, but when the bayonet attachment is assembled to the riot type of these guns, the barrel cannot be removed unless the bayonet attachment is first removed. Therefore, these guns, with bayonet attachment assembled, are considered as solid-frame guns. In the case of solid-frame guns as explained above, the bore should be cleaned from the muzzle end without removing bayonet attachment or barrel, unless it is necessary to remove the groups from the receiver for periodic cleaning or when the gun is exposed to extreme conditions as explained in this manual.

j. Disassembly and assembly, as treated in this technical manual, comprise the removal and replacement of only such groups of parts as are necessary to a thorough cleaning of the gun. Should further disassembly be necessary for adjustment or repair, the gun should be turned over to ordnance personnel. A group is a number of parts which either function together or are intimately related to each other and should, therefore, be considered together. A group may be composed of one assembly of two or more parts or sub-assemblies, or a number of assemblies and parts. For example, the barrel, magazine, and action slide group is composed of the barrel assembly, the magazine assembly and the action slide assembly. In removing the groups or parts from the gun, it is often necessary to remove other groups or parts first, in order to be able to remove the group or part desired. In most instances the groups or parts must be removed and replaced in the order and manner prescribed herein.

k. Modern shotguns may be roughly divided into five general classes: single-barrel, double-barrel, manually operated repeaters, autoloading repeaters, and multi-barrel.

SHOTGUNS, ALL TYPES

(1) Single barrel guns have a single barrel only and are thus capable of firing but one shot without reloading. Such guns may be of the hammer or hammerless (usually) design and are manually operated.

(2) Double-barrel guns have two barrels, mounted either side-by-side or one under the other. In the latter case they are termed "over-and-under," or "over-under" guns. The side-by-side design, usually termed "double-barrel," may be hammer or hammerless (usually) guns, while the "over-under" design is usually hammerless. Both designs are manually operated and are capable of firing two shots without reloading.

(3) Repeating guns have a single barrel with a tubular magazine below it, and are capable of firing from 3 to 6 shots without reloading. Such guns are manually operated by means of a sliding action which loads, cocks, and clears the gun when operated. The slide action guns covered in this manual are of this design.

(4) Autoloading repeaters have a single barrel with a tubular magazine below it and are capable of firing from 3 to 5 shots without reloading. Such guns must be loaded manually for the first shot. When the gun is fired, the recoil operates the mechanism to clear, cock, and load the gun from the magazine. The only operation necessary for the operator to perform after the initial loading and cocking is to pull the trigger. The autoloading guns covered in this manual are of this design.

(5) Multi-barrel guns have three or more barrels. Usually they have two shotgun barrels mounted side by side with a rifle barrel below them. Such guns are usually of the hammerless design and are manually operated.

l. Shotguns produced by various manufacturers, are usually referred to as of a specific model and grade. The model refers to the basic mechanical design of the gun; the grade, to the superficial design such as engraving, special stock, ribbed barrel, or modifications of design. Grades are referred to herein only when there is a variation in mechanical design. The various grades of a specific model may come in different gages, length of barrels, or degrees of choke in barrel boring. A sporting skeet gun, however, is usually furnished with a 26-inch barrel and a sporting trap gun with a 30- or 32-inch barrel. Guns used for game shooting may come with any length barrel or degree of choke. Such guns usually have a plain barrel without a rib. In this technical manual, however, due to the fact that guns of various grades are being issued, guns are classed as three types, riot, sporting skeet, and sporting trap, according to length of barrel and degree of choke only. Such guns may be of any grade. Therefore any gun with a 20-inch cylinder barrel is classed as a riot gun; any gun with a 26-inch, improved cylinder barrel as a sporting skeet gun, and any gun with a 30-inch, full choke barrel as a sporting trap gun. Riot guns are specified to be furnished with bayonets, hand guards and sling swivels,

INTRODUCTION

but some have been procured without these accessories, in the interests of expediency. These are termed "substitute riot guns." Sporting guns as referred to herein are used principally for trapshooting.

NOTE: The Migratory Game Act provides that not more than three shells shall be contained in a repeating or autoloading shotgun, and that the magazines shall therefore be limited to a capacity of two shells. To comply with this act, a hardwood plug is furnished for insertion in the end of the magazine to reduce its capacity to two shells.

m. The shotgun shell or cartridge is cylindrical in shape and may be composed entirely of brass, or a paper casing seated in a brass or steel base. The primer is seated in the base of the shell, which contains the powder charge and shot pellets, separated and held in position by cylindrical wads of felt and cardboard. The primer, when struck by the firing pin, detonates and produces a spark which ignites the powder. The expansion of the gas generated by the burning powder furnishes the force for propelling the shot charge from the gun. For description of the shell refer to section XIII.

n. In addition to field inspection as prescribed for each gun, the barrel should be inspected, and the trigger pull tested as follows:

(1) INSPECTION OF BARREL. The barrel should be inspected for looseness in receiver, rust, pits, leading, cracks, and bulges.

(a) If barrel is loose (shakes) in receiver, the gun should be turned over to ordnance personnel for correction.

(b) Rust and leading may be removed if not too bad (par. 11 d). If barrel is badly rusted or pitted, the gun should be turned over to ordnance personnel.

(c) If cracks or bulges are evident, the gun should be turned over to ordnance personnel. A bulge is usually indicated by a shadowy depression or dark ring in the bore, and may often be noticed through a bulge or raised ring on the barrel surface.

(2) TRIGGER PULL.

(a) Trigger pull should in general range between 5 pounds (minimum) and 8 pounds (maximum) for the guns covered in this bulletin. Trigger pull can be tested with the regulation trigger pull test weight hook and weights used for rifles and carried in the small arms repair truck, or hook and weights improvised as in (b) below.

(b) The inspector, in testing trigger pull of shotguns, should have two weights, one of 5 pounds and one of 8 pounds. Each of the weights should be provided with a wire so that the pressure will be applied ¼ inch from the lower end of the trigger and exerted parallel with the axis of the bore. The wire should be stiff enough to retain an L-hook bend,

SHOTGUNS, ALL TYPES

not less than 2¼ inches long, in the free end, and long enough to allow the weight to swing clear of the butt end of the stock when testing.

(c) To test the trigger pull, note that the gun is fully unloaded, action locked, hammer fully cocked, and the safety set to the fire position thus allowing the trigger to be retracted. Have the weight resting on the floor or ground, and insert hook of trigger weight wire (or test hook) through the trigger guard bow to bear on the trigger, so that the pressure will be applied ¼ inch from the lower end of the trigger. Care should be taken during the test to see that the wire contacts the trigger only and does not rub against the trigger guard bow or stock, and that wire and axis of bore are parallel and perpendicular. Then with the barrel of the gun held vertically, raise the weight from the floor as gently as possible. If the 5-pound weight pulls the trigger, or the 8-pound weight fails to pull the trigger, the gun should be turned over to ordnance personnel.

o. The illustrations in this technical manual pertaining to the operation of the gun are for explanatory purposes only. The current regulations or manual of arms, on the manner in which the gun should be held while operating, should be followed.

p. The word "breech" refers to the section of the gun just to the rear of the barrel chamber, where the gun is loaded. The word "muzzle" refers to the extreme forward end of the barrel.

q. **Safety Precautions.**

(1) Every shotgun should be considered to be fully loaded and cocked until it has been personally examined by the operator and proved to be otherwise. Memory should never be trusted as to a gun's condition in this respect.

(2) A shotgun should never be pointed at anyone at whom it is not intended to shoot, nor in a direction where accidental discharge may do harm.

(3) A shotgun should always be fully unloaded if it is to be left where someone else may handle it.

(4) A shotgun should always be pointed up to a safe spot when pulling the trigger after examination.

(5) If a shotgun is to be carried cocked, with a shell in the chamber, the trigger should be blocked by sliding the safety to the safe position. In the case of the Winchester Gun M97, the hammer should be placed at half-cock.

(6) Under no circumstances should pressure be applied to the trigger while the gun is being operated, until the breech bolt (or like part) is positively locked, and the slide (or like part) blocked.

(7) Under no circumstances should the hammer of the Winchester Gun M97 be let completely down with a shell in the chamber.

INTRODUCTION

(8) A shotgun should never be fired with any grease, cleaning patch, dust, dirt, mud, snow, or other obstruction in the bore. To do so may burst the barrel or blow the bolt.

(9) Ammunition should never be greased or oiled. This will affect the ammunition, and creates a hazardous pressure on the bolt.

(10) Chamber and bore should be wiped dry of oil or grease before firing for the reason given in (8) and (9) above.

(11) Ammunition should be clean and dry; all live and dummy ammunition should be carefully examined; all defective, swollen, or badly bruised shells turned in.

(12) Ammunition should not be exposed to the direct rays of the sun for any length of time. This increases chamber pressure and affects the charge (par. 92).

(13) The bore should always be inspected before loading the gun.

(14) A gun presumed to be fully unloaded should never be handed to anyone until again inspected.

(15) Shotgun shells smaller than 12-gage should never be carried. A smaller shell, if accidentally loaded, will enter the bore and burst the barrel when the gun is fired.

CAUTION: The proper functioning of the bolt and/or slide locking mechanisms of the guns covered in this manual is of the utmost importance to the proper operation and functioning of the gun, and the safety of the operator. If a bolt fails to lock properly, or a slide lock (or similar part) fails to function to prevent premature unlocking of the bolt, the breech is apt to blow open when the gun is fired with possible injury to the operator. Extreme care should be observed to see that such parts and those which operate and function them are in good repair and adjustment at all times. Care should be observed by operators to see that the breech mechanism is securely locked and blocked as directed in the field inspection for the gun in question when operating the guns.

SHOTGUNS, ALL TYPES

Section II

WINCHESTER SHOTGUN, 12-GAGE, M97

	Paragraph
Description	4
Data	5
Operation	6
Functioning	7
Removal of groups	8
Replacement of groups	9
Field inspection	10
Cleaning and lubrication	11

4. DESCRIPTION.

a. Identification marks on this gun are generally to be found as follows:

(1) Name of maker, gage, model, and barrel boring are stamped on the top of the barrel near the breech end.

(2) Serial number of the gun is stamped on the lower face of the barrel near the breech end, and on the lower face of the forward end of the receiver.

(3) On solid-frame gun, maker's name and model are stamped on the action slide bar, and the serial number of the gun as in (2) above.

b. This gun (figs. 1, 2, 3, and 4) is a manually operated repeating shotgun of the slide action, hammer type, manufactured in both solid-frame and take-down types. In the solid-frame gun (figs. 1 and 2) the barrel is screwed into the receiver at manufacture and is not intended to be removed except for repair; the magazine is also screwed into the receiver but can easily be removed by removal of a stop screw and unscrewing the magazine from the receiver. The take-down gun is so constructed that the barrel and magazine together with the action slide group can easily be removed as a unit by disengaging interrupted threads on rear end of the magazine and barrel from like interrupted threads in the receiver by which the barrel and magazine are locked to the receiver. This construction facilitates cleaning and transportation of the gun.

c. Riot guns of early manufacture were of the solid-frame design, while such guns of later manufacture were of a modified solid-frame design. In this later design the take-down type of receiver, barrel and magazine (with action slide assembled) were used. The barrel and magazine (with action slide assembled), however, must be removed separately from the receiver, and not as a unit as in the take-down gun. This modified gun has a receiver extension screwed to the barrel similar to the take-down

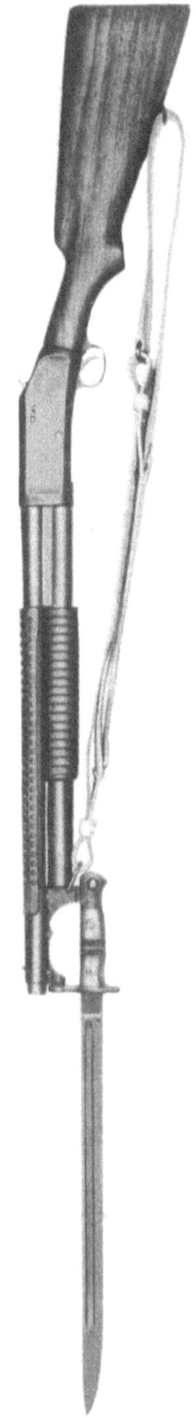

TM 9-285
4

WINCHESTER SHOTGUN, 12-GAGE, M97

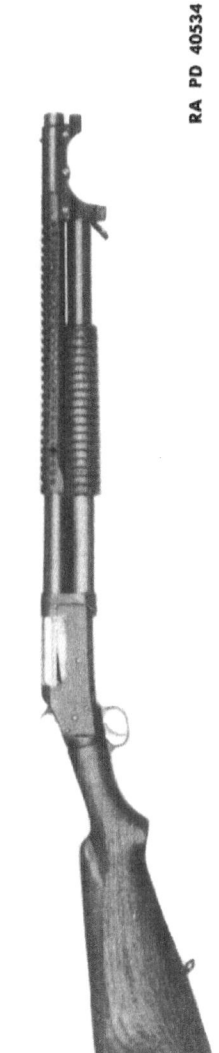

RA PD 40535

RA PD 40534

Figure 1 — Gun with Bayonet, Hand Guard and Sling — Left Side View — Riot Type (Solid-Frame) — Winchester Shotgun M97

Figure 2 — Gun with Bayonet Attachment and Hand Guard — Right Side View — Riot Type (Solid Frame) — Winchester Shotgun M97

SHOTGUNS, ALL TYPES

Figure 3—Left Side View—Sporting Trap Type (Take-down)—Winchester Shotgun M97

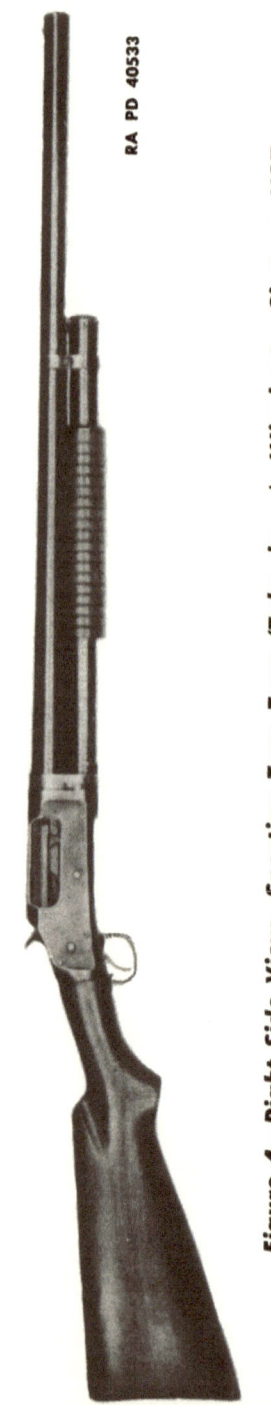

Figure 4—Right Side View—Sporting Trap Type (Take-down)—Winchester Shotgun M97

WINCHESTER SHOTGUN, 12-GAGE, M97

gun, while the barrel of the solid-frame gun of early manufacture, has no such extension, and screws directly into the receiver. The magazine band is absent in both guns when the bayonet attachment is assembled to the gun, and the magazine plug has an integral stud on its forward end. Barrel and magazine of the two guns are not interchangeable (par. 3 i).

d. The take-down design of this gun is furnished in various grades having barrels of different lengths and degrees of boring and other modifications of design. Basically, however, the mechanism of the guns is identical except for the differences in design mentioned in h and c above. For convenience herein the guns will be classified as three types: riot, sporting skeet, and sporting trap, although variations of these types may occur.

(1) The riot-type gun (figs. 1 and 2) may come in either the solid-frame or take-down design, with a 20-inch plain barrel, bored cylinder. Most of these guns have a bayonet attachment and hand guard (fig. 5) attached to the muzzle end of the barrel and the magazine, and a leather sling attached to a sling swivel on the bayonet attachment and the stock. These are always of the solid-frame or modified solid-frame type.

(2) The sporting skeet-type gun usually comes in the take-down design only and is furnished with a 26-inch plain or ribbed barrel, bored improved cylinder, and is without bayonet attachment or sling.

(3) In the sporting trap-type gun (figs. 3 and 4) is similar to the skeet gun but is made with a 30-inch barrel, bored full choke, and is without bayonet attachment or sling.

e. **General Description** (figs. 6, 7, 8, 14, and 15).

(1) The stock of the gun is bolted to the rear end of the receiver, and the barrel and magazine are fastened to the forward end of the receiver in either of the ways explained in h and c above. The action slide is mounted and operates on the magazine. The rear end of the slide or bar passes through the forward end of the receiver and engages with and cam-operates the carrier, pivoted in the receiver, and at the same time reciprocates the breech bolt through the medium of a hook pivoted on the bolt.

(2) The receiver contains the operating mechanism and to its lower rear end is attached the trigger plate in which the trigger is mounted. The receiver is open at the bottom to permit loading, the rear to permit rearward passage of the breech bolt, and the right side for ejection of the fired shell cases.

(3) The breech bolt (fig. 8) contains the firing pin, firing pin lock and the extractors. The carrier (fig. 8) contains the hammer, sear, and action slide lock, together with their springs and components. The trigger plate contains the trigger, trigger spring, and stop screw. The ejector is mounted to the left wall of the receiver and the action slide lock release plunger pin in the right wall of the receiver.

TM 9-285

SHOTGUNS, ALL TYPES

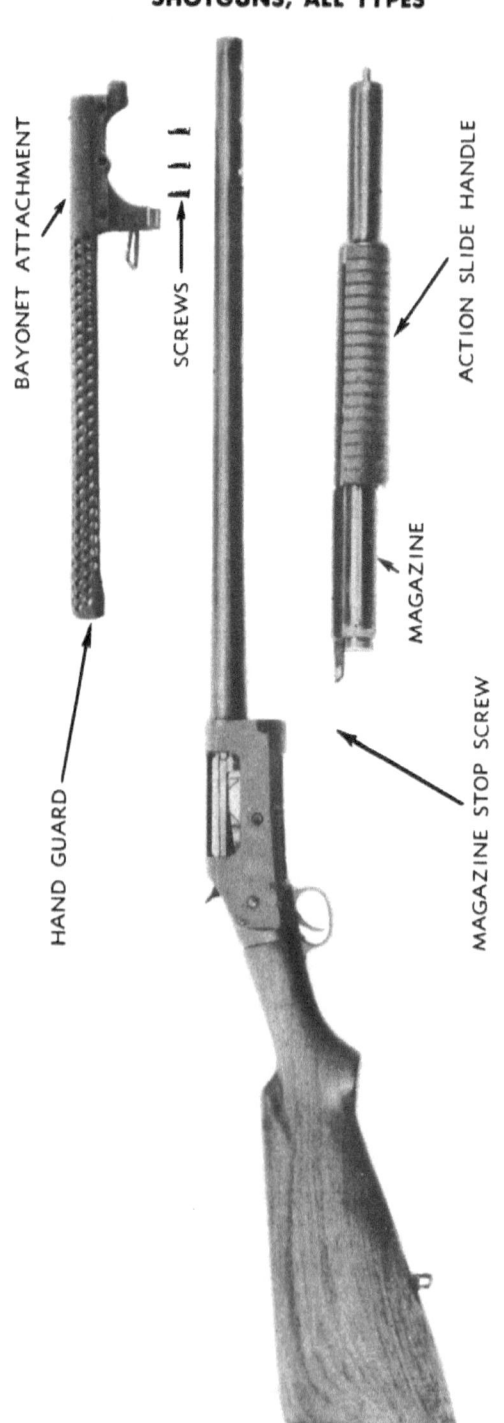

Figure 5—Bayonet Attachment, Magazine and Action Slide Groups Disassembled from Gun—Riot Type (Solid-Frame)—Winchester Shotgun M97

TM 9-285
4

WINCHESTER SHOTGUN, 12-GAGE, M97

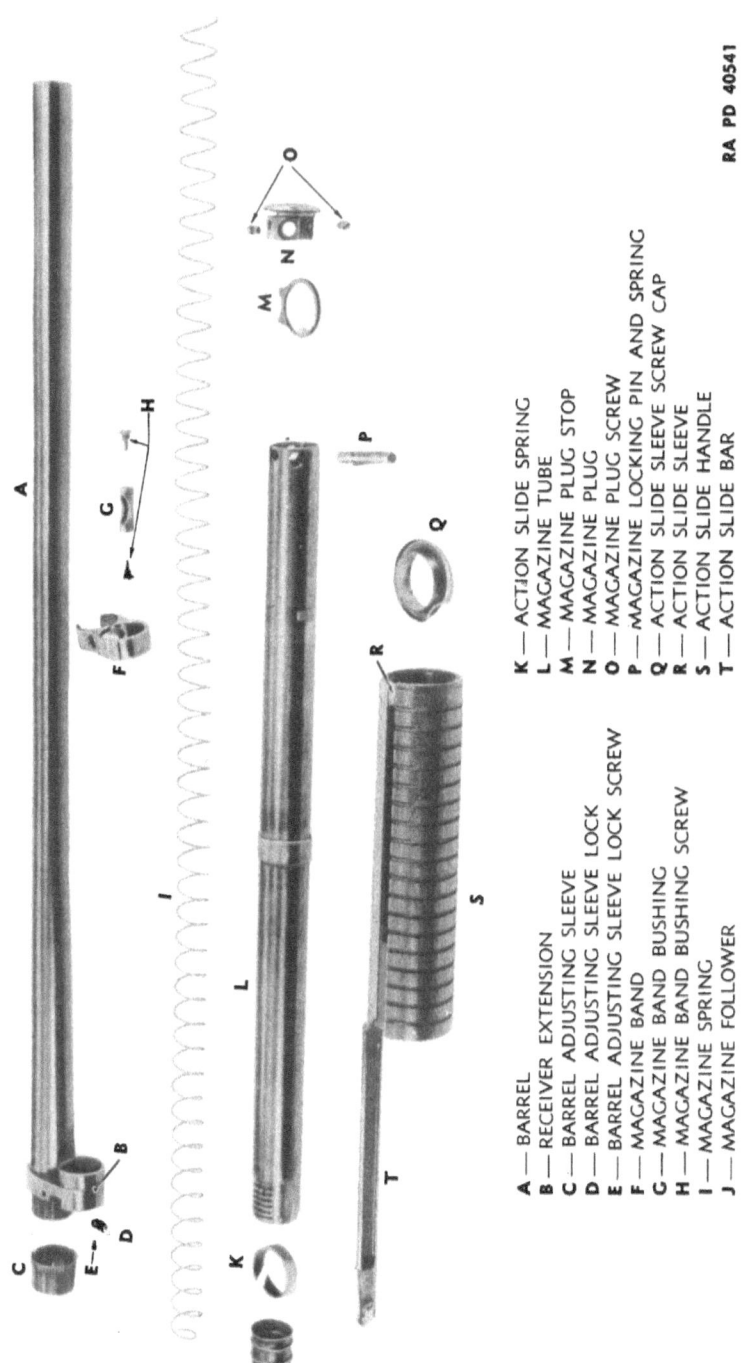

A — BARREL
B — RECEIVER EXTENSION
C — BARREL ADJUSTING SLEEVE
D — BARREL ADJUSTING SLEEVE LOCK
E — BARREL ADJUSTING SLEEVE LOCK SCREW
F — MAGAZINE BAND
G — MAGAZINE BAND BUSHING
H — MAGAZINE BAND BUSHING SCREW
I — MAGAZINE SPRING
J — MAGAZINE FOLLOWER
K — ACTION SLIDE SPRING
L — MAGAZINE TUBE
M — MAGAZINE PLUG STOP
N — MAGAZINE PLUG
O — MAGAZINE PLUG SCREW
P — MAGAZINE LOCKING PIN AND SPRING
Q — ACTION SLIDE SLEEVE SCREW CAP
R — ACTION SLIDE SLEEVE
S — ACTION SLIDE HANDLE
T — ACTION SLIDE BAR

Figure 6 — Barrel, Magazine and Action Slide Group — Disassembled View — Winchester Shotgun M97 — (Take-down)

SHOTGUNS, ALL TYPES

(4) There is no trigger safety on this gun, this function being performed by setting the hammer at the half cock which prevents retraction of the trigger and release of the action slide bar.

(5) The magazine which is of the tubular type is positioned beneath the barrel and has a capacity of five shells loaded end to end. The shells are pressed together and fed into the receiver by the force of the magazine spring acting upon the follower.

(6) The shell stops are pivoted in the lower, forward inside walls of the receiver, and act to hold the shells in the magazine against the pressure of the magazine spring. The stops are operated by the carrier to release and block the shells at the proper time, thus allowing but one shell to enter the receiver at a time.

(7) The action slide lock (figs. 12 and 13) is positioned in the left side of the carrier and acts upon the action slide bar to block its rearward movement after the breech bolt is locked in position by the carrier, thereby preventing premature unlocking of the bolt. The lock is disengaged either by the descending hammer pressing on the action slide lock release plunger, when the gun is fired, or by manually pressing on the action slide lock release plunger pin. When the lock is disengaged from the rear end of the action slide bar the action slide may be moved to the rear to cam down the carrier, unlock the bolt and function the operating mechanism.

(8) The bayonet attachment (fig. 5) and hand guard common to the riot-type gun only are riveted together and mounted on the muzzle end of the barrel and forward end of the magazine, and held in place by means of a stud on the magazine plug, and screws passing through the attachment and grooves in the barrel. A sling swivel is attached to the attachment and stock respectively, and supplies the means for fastening the sling to the gun.

(9) The gun sling is of leather of the M1907 model and the bayonet and bayonet scabbard are the M1917 model (figs. 1 and 2).

5. DATA.

Gage of bore	12
Boring of barrel—riot type	Cylinder
Boring of barrel—sporting skeet type	Improved cylinder
Boring of barrel—sporting trap type	Full choke
Type of action	Slide
Type of firing mechanism	Hammer
Type of magazine	Tubular
Capacity of magazine	5 shells
Length of barrel—riot type	20 in.
Length of barrel—sporting skeet type	26 in.

WINCHESTER SHOTGUN, 12-GAGE, M97

Length of barrel—sporting trap type . . . . 30 in.
Length of stock and receiver (approx.) . . . . 19½ in.
Length of assembled gun—riot type (approx.) . . . . 39 in.
Length of assembled gun—sporting skeet type (approx.) . . . . 45 in.
Length of assembled gun—sporting trap type (approx.) . . . . 49 in.
Weight of assembled gun—riot type (approx.) . . . . 8 lb
Weight of assembled gun—sporting skeet type (approx.) . . . . 7⅜ lb
Weight of assembled gun—sporting trap type (approx.) . . . . 7⅝ lb
Weight of Bayonet M1917 (approx.) . . . . 1⅛ lb

6. OPERATION.

a. The gun is operated by moving the action slide handle smartly and fully, backward and forward. This operation unlocks the breech bolt, extracts and ejects the fired shell, cocks the hammer, transfers a live shell from the magazine to the chamber of the barrel and relocks the breech bolt behind the shell.

CAUTION: During operation the muzzle of the gun should always be pointed at a safe spot.

b. Before the action slide can be retracted, the action slide lock must be disengaged from the action slide bar (fig. 9). If the gun has been fired, and the hammer consequently forward, the only movement necessary is to move the action slide handle forward slightly to allow the lock to disengage, and then reciprocate it as above, as the descending hammer has already partly disengaged the lock. If the hammer is at half cock, it must be pulled back to full cock, and the action slide lock release plunger pin then pressed and the handle pushed forward and reciprocated as explained. If the gun is at full cock, proceed as just stated.

CAUTION: During these operations, the finger should remain outside the trigger guard. Reciprocation of the slide handle should be full and smart to insure extraction of the shell, cocking of the hammer, complete locking of the breech bolt, and blocking of the action slide. Slamming of the mechanism, however, should be avoided. When the gun is being fired as a repeater, all pressure should be removed from trigger while operating.

c. When the gun is being fired as a repeater, the recoil of the gun performs the preliminary forward movement of the action slide handle, as the gun recoils away from the handle which is held by the operator.

d. With the gun loaded and locked, and the hammer at full cock, the only operation necessary to fire the gun is to pull the trigger to release the hammer.

e. **To Load and Unload the Magazine (fig. 11).**

(1) To load the magazine, press a shell nose first, into the rear of

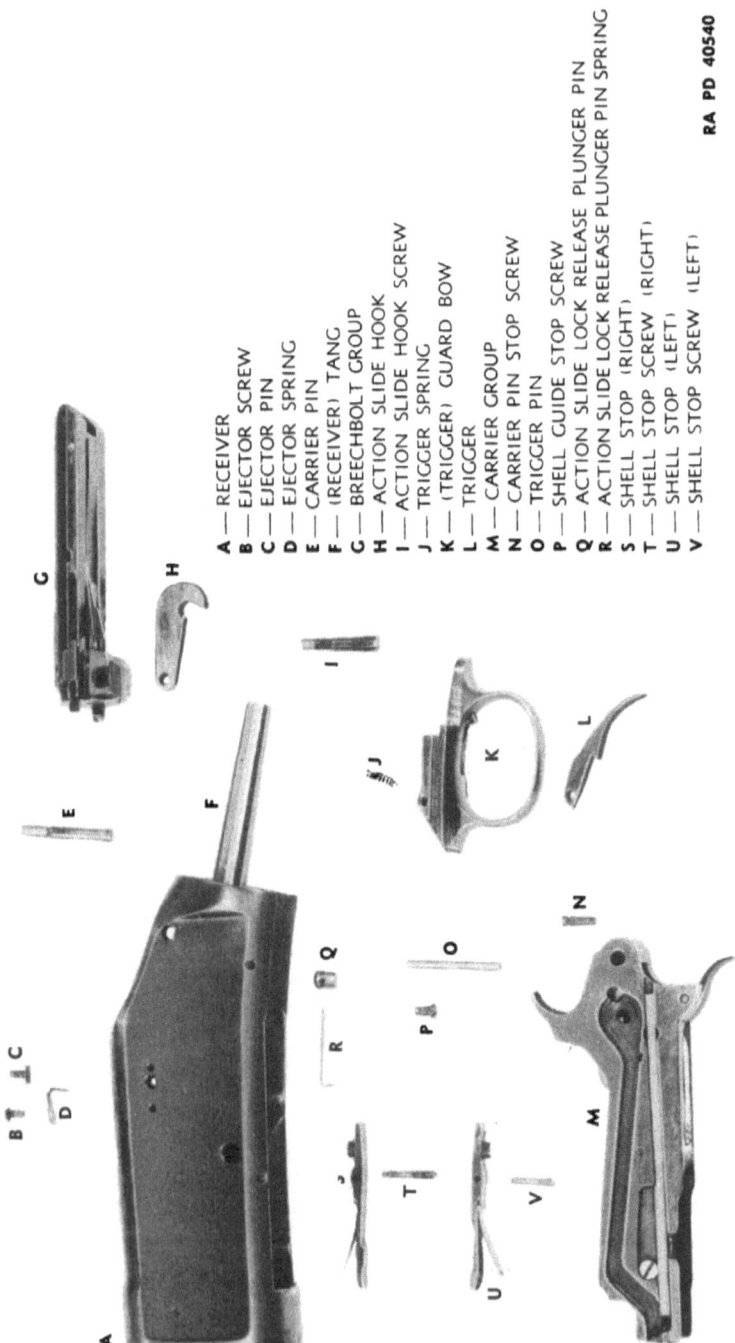

Figure 7 — Receiver Group — Disassembled View — Winchester Shotgun M97

TM 9-285

WINCHESTER SHOTGUN, 12-GAGE, M97

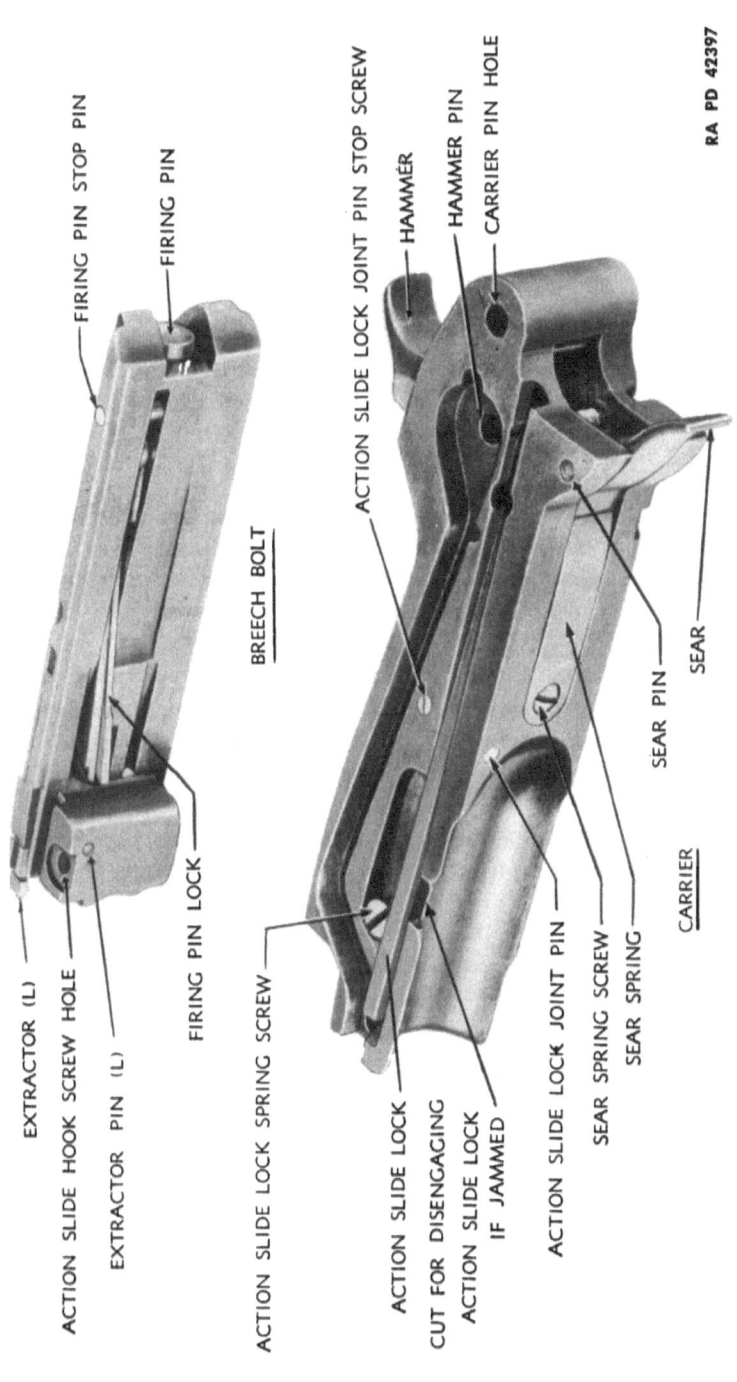

Figure 8 — Breech Bolt and Carrier — Rear, Bottom and Left Side View — Showing Location of Parts — Winchester Shotgun M97

SHOTGUNS, ALL TYPES

the magazine against the magazine follower until the shell stops at the mouth of the magazine, snap out behind and retain the shell. Load another in the same way by pressing it against the base of the first shell loaded, until five in all are loaded. Loading should be done with breech bolt locked and hammer at half cock.

(2) To unload the magazine, place hammer at half cock (fig. 10), press in the two shell stop plungers projecting through the walls of the receiver, and allow the shells to run out of the magazine. Inspect magazine to see that it is empty. Then retract action slide handle to eject shell in the chamber and inspect chamber and receiver to be sure gun is empty.

CAUTION: Never let hammer fully down when there is a live shell in the chamber. To release action slide lock, pull hammer back to full cock and, with fingers outside trigger guard, press lock release plunger pin and retract slide as directed in h above.

f. To Load and Unload the Gun.

(1) To load a shell from the magazine into the chamber, pull back hammer to full cock, press action slide lock release plunger pin, and reciprocate action slide as explained. Then let hammer down to half cock (safe) position. Another shell may then be loaded into the magazine.

(2) To load the chamber only with magazine empty, retract action slide handle, place shell directly in chamber of barrel through ejection opening in receiver, lock breech bolt, and place hammer at half cock.

(3) To unload the gun, place hammer at half cock, unload magazine; then place hammer at full cock, press action slide lock release plunger pin, and retract action slide handle to extract and eject shell in chamber. Then inspect magazine and chamber to be sure gun is completely unloaded.

CAUTION: To let hammer down from full to half cock, hold hammer firmly with thumb, press trigger, and ease hammer down slightly beyond half cock position; then with pressure on trigger removed, pull hammer back until it definitely clicks into position at half cock before releasing. Release trigger as soon as hammer is released from full cock.

7. FUNCTIONING.

a. As already briefly explained, the functioning of the operating mechanism is accomplished by the reciprocation of the action slide handle. A cam lug on the rear end of the action slide bar engages in an irregular camming aperture in the left side of the carrier and as the slide is moved backward, the carrier is cammed down; as the slide is moved forward, the carrier is cammed up.

h. The action slide hook, pivoted on the left side of the breech bolt

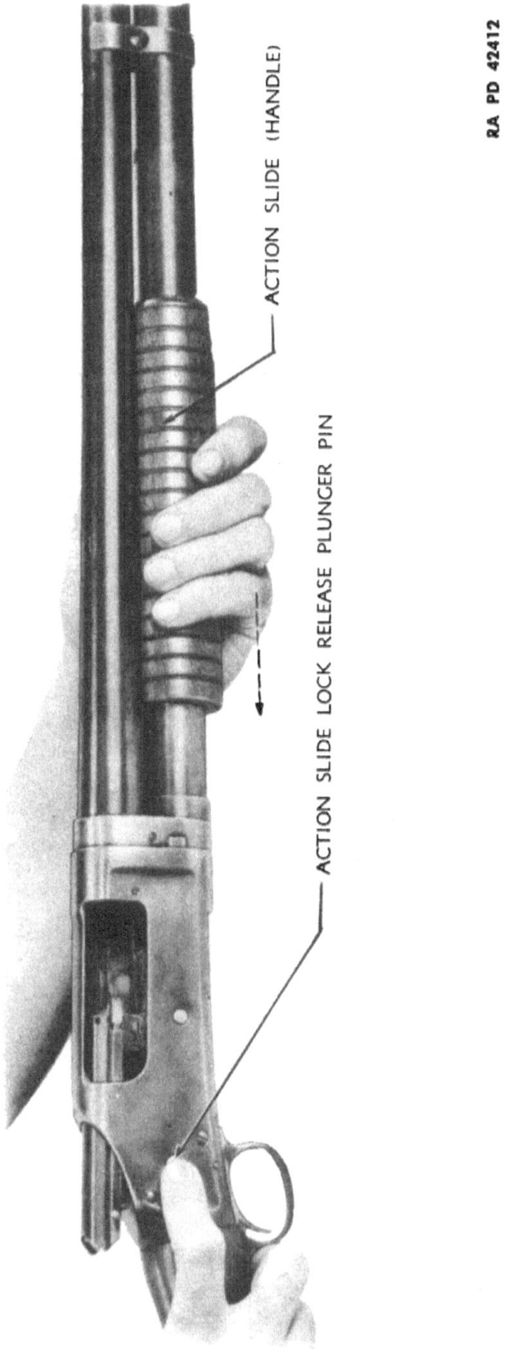

WINCHESTER SHOTGUN, 12-GAGE, M97

Figure 9 — Depressing Action Slide Lock Release Plunger Pin and Retracting Action Slide — Winchester Shotgun M97

TM 9-285
7

SHOTGUNS, ALL TYPES

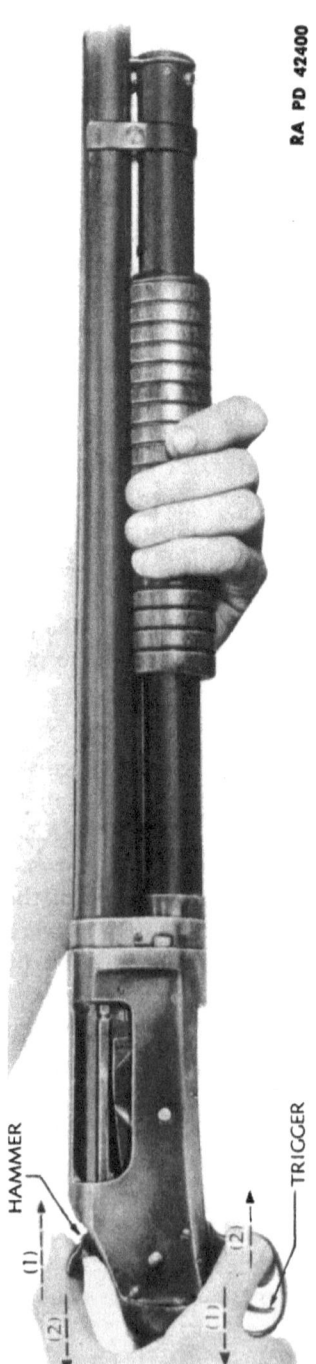

Figure 10 — Setting Hammer at Half Cock (Safe) Position — Winchester Shotgun M97

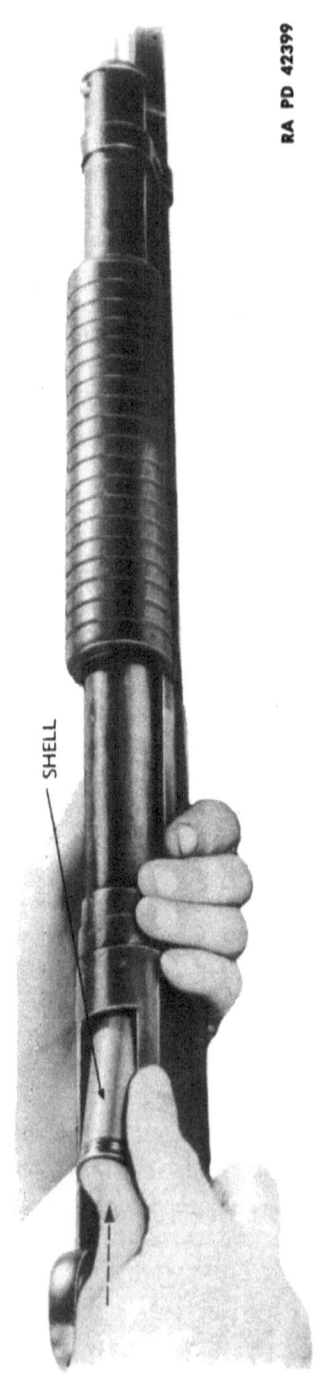

Figure 11 — Loading Magazine — Winchester Shotgun M97

WINCHESTER SHOTGUN, 12-GAGE, M97

(figs. 12 and 13), engages with the rear end of the action slide bar, and as the bar is reciprocated, the bolt is likewise moved back and forth. The timing of the movements of the carrier and bolt is such that the carrier is cammed completely up as the bolt reaches the extreme forward position. The locking shoulder on the forward end of the carrier is thus moved up behind the locking shoulder on the forward end of the bolt, locking the bolt in the closed position. As the bolt is locked in position, the action slide lock, pivoted in the left side of the carrier, springs out to block the rear end of the action slide bar (fig. 13). The action slide is thus prevented from moving to the rear and unlocking the bolt until the lock is disengaged either by the descending hammer, when the gun is fired, or by manual depression of the action slide lock release plunger pin extending through the right side of the receiver.

c. As the breech bolt moves to the rear, it cams back and cocks the hammer, which is caught and held by the sear, positioned in the rear end of the carrier. The hammer is released by pulling the trigger mounted below it, which levers the sear from engagement with the sear notch in the hammer. The firing pin is cammed back into the breech bolt by the firing pin lock as the carrier unlocks the breech bolt. The pin is held thus retracted until the bolt is again locked, at which time it is released by a projection on the carrier pressing on, and disengaging the lock to free the firing pin.

d. The carrier depresses the shell stops, positioned in the receiver directly behind the magazine, on its downward movement, thus releasing a shell from the magazine and at the same time takes over the function of the stops by blocking the next shell in the magazine. The released shell is pushed into the receiver, upon the carrier, by the force of the magazine spring and lifted to chamber level by the carrier in its upward movement. The breech bolt pushes the shell into the chamber on its forward movement and is locked behind it as already explained. As the carrier rises to lock the bolt, it releases the shell stops which spring outward to block the shell the carrier has been blocking. The shell guide, pivoted on the carrier, is cam-operated by the head of the action slide hook screw, projecting from the right side of the bolt, to clear the ejection opening in the receiver when the bolt is retracted.

e. The shell is extracted from the chamber by the extractors, positioned in the forward end of the breech bolt as the bolt moves to the rear, and knocked out through the ejection opening in the right side of the receiver by striking the ejector, positioned in the left inner wall of the receiver.

8. REMOVAL OF GROUPS (figs. 6, 7, 14, and 15).

a. Groups and parts should be removed and replaced in the order

TM 9-285

SHOTGUNS, ALL TYPES

Figure 12 — Breech Bolt, Carrier and Action Slide Bar Groups — Breech Bolt Unlocked from Carrier — Action Slide Lock Disengaged — Winchester Shotgun M97

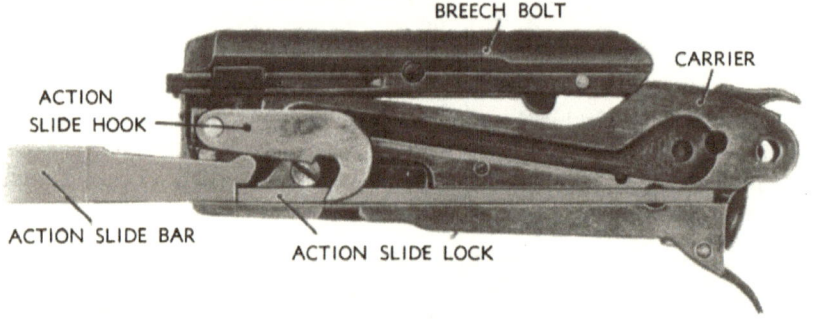

Figure 13 — Breech Bolt, Carrier and Action Slide Bar Groups — Breech Bolt Locked by Carrier — Action Slide Lock Engaged — Winchester Shotgun M97

given below. In the take-down design of this gun, it is necessary to remove the barrel, magazine, and action slide group before removing the groups from the receiver. In the solid-frame design gun, the bayonet attachment, magazine, and action slide group must be removed. The barrel is screwed into the receiver at manufacturer and should not be removed. The other groups, contained in the receiver of these two guns, are practically the same and can be removed in the same manner. Groups and parts when removed should be placed on a clean, flat surface and care observed to prevent loss of screws and small parts. Remove as follows:

(1) BAYONET ATTACHMENT, MAGAZINE, AND ACTION SLIDE GROUP (SOLID-FRAME GUN).

(a) Bayonet Attachment. Unscrew and remove the three transverse

WINCHESTER SHOTGUN, 12-GAGE, M97

screws from the bayonet attachment and drive forward off barrel. The barrel of the solid-frame gun of early manufacture is screwed into the receiver at manufacture, and should not be removed (par. 8 (1)(c)).

(b) *Magazine*. Lock action by pushing action slide handle fully forward, and remove magazine stop screw from lower, forward end of right side of receiver. Then using improvised tool with pin and hole in end, or rod bent to an L, to engage stud and hole in end of magazine plug, turn magazine counterclockwise out of receiver and withdraw with action slide group attached.

(c) The barrel and magazine of the modified type of solid-frame gun (par. 4 c) may be removed separately but should not be removed unless necessary. The magazine is first removed (par. 8 a (1)(b)) by turning ¼ turn clockwise to disengage the interrupted threads on magazine from those in receiver and then withdrawing magazine and action slide together from receiver. The barrel can then be removed by turning ¼ turn clockwise and withdrawing from receiver. This gun can be identified by the receiver extension screwed to the rear end of the barrel, similar to the take-down gun.

(2) BARREL, MAGAZINE, AND ACTION SLIDE GROUPS (TAKE-DOWN GUN).

(a) Lock action by pushing action slide fully forward. Unlock the magazine by pressing the locking pin through the plug until it will clear the barrel when the magazine is turned (fig. 16).

(b) Turn the magazine ¼ turn clockwise, using the pin as a lever, to disengage the interrupted threads on the rear end of the magazine from those in the receiver. Then, pull the magazine forward out of engagement with the receiver.

(c) Push the action slide handle forward until the rear end of the bar is free of the receiver. Then, grasp barrel and magazine in left hand and receiver in right hand, and turn the barrel ¼ turn clockwise to disengage the interrupted threads on the rear of the barrel from those in the receiver, and withdraw the group forward from the receiver (fig. 17).

(3) CARRIER GROUP. With barrel, magazine, and action slide groups (take-down gun) or magazine and action slide group (solid-frame gun) removed, the carrier may be removed as follows:

(a) Unscrew the carrier pin stop screw which is visible in the left top wall of the carrier just to the rear of the hammer tang. With this screw removed, drift out the carrier pin extending through both receiver walls just below the carrier seat. Next, remove the shell guide stop screw located at the lower rear extremity of the right-hand wall of the receiver.

(b) With the hammer cocked, depress the action slide lock release plunger pin, extending through the right wall of the receiver, to prevent

TM 9-285

SHOTGUNS, ALL TYPES

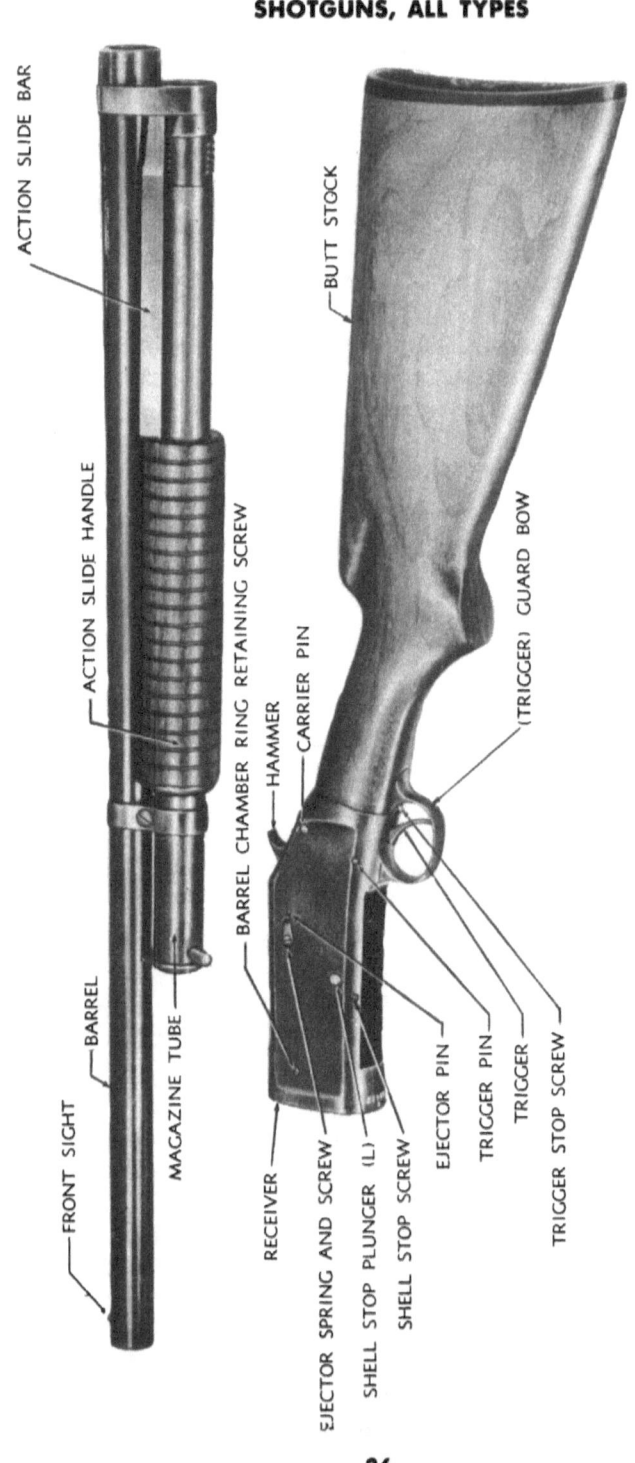

Figure 14—Gun Taken Down—Left Side View—Showing Location of Parts—Winchester Shotgun M97 (Take-down)

TM 9-285
8

WINCHESTER SHOTGUN, 12-GAGE, M97

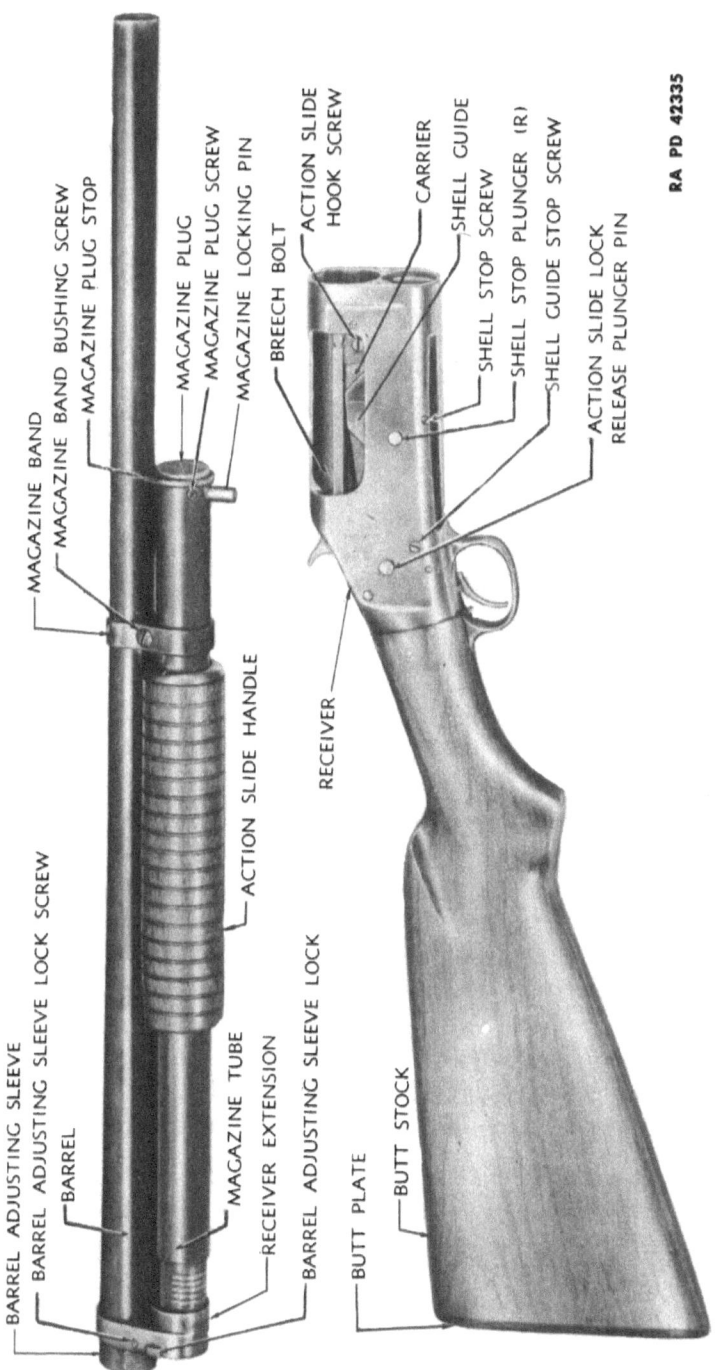

Figure 15—Gun Taken Down—Right Side View—Showing Location of Parts—Winchester Shotgun M97 (Take-down)

SHOTGUNS, ALL TYPES

interference of lock with receiver wall. Then, insert a screwdriver through the ejection opening and pry gently on the carrier to start it downward, after which it can be rotated down and out of the receiver through the lower opening.

NOTE: If release plunger will not function, the forward end of the lock can be depressed by inserting a small screwdriver blade into the cut in the lower left forward face of the carrier and prying the lock inward.

(4) BREECH BOLT GROUPS. With the carrier removed, the breech bolt may be removed as follows:

(a) With the breech bolt fully forward, remove the action slide hook screw from the right forward face of the breech bolt.

(b) Reach into the receiver and remove the action slide hook from the left side of the breech bolt. The hook is easily removed, after the screw has been removed.

(c) Slide the breech bolt group to the rear out of the receiver.

(5) TRIGGER GUARD BOW GROUP. Ordinarily it should not be necessary to remove the guard bow group for cleaning. To remove this group, the butt stock must first be removed. This can be accomplished by removing butt plate screws and butt plate. Then with long shanked screwdriver disengage butt stock bolt now visible in hole in stock, from tang of receiver by turning counterclockwise, and then pull butt stock to rear from receiver. The guard bow group can then be removed as follows:

(a) With pin drift, drive out trigger pin from lower rear end of receiver and pull guard bow to the rear out of receiver.

(b) Trigger and trigger spring can then be removed from guard bow. The stop screw should not be removed. This screw is used to limit rearward movement of trigger, which should move far enough to disengage sear from hammer when retracted.

9. REPLACEMENT OF GROUPS.

a. Groups and parts should be thoroughly cleaned, lightly oiled and lubricated, if necessary, before replacing. Replace as follows:

(1) TRIGGER GUARD BOW GROUP.

(a) Seat trigger spring in its aperture in guard bow, and place trigger on top of spring with curved end extending downward through floor of guard bow.

(b) Slide guard bow into rear of receiver, with pin hole forward, until pin hole in guard bow and receiver are in alinement.

(c) Press trigger down against spring until pin hole in trigger alines

WINCHESTER SHOTGUN, 12-GAGE, M97

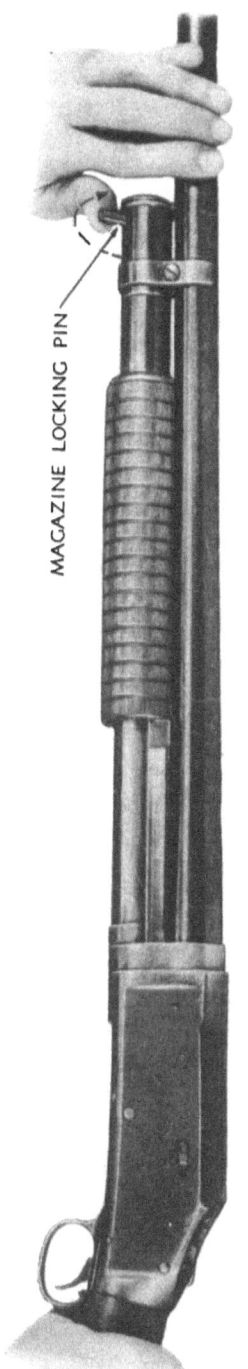

Figure 16 — Unlocking Magazine from Receiver — Winchester Shotgun M97 (Take-down)

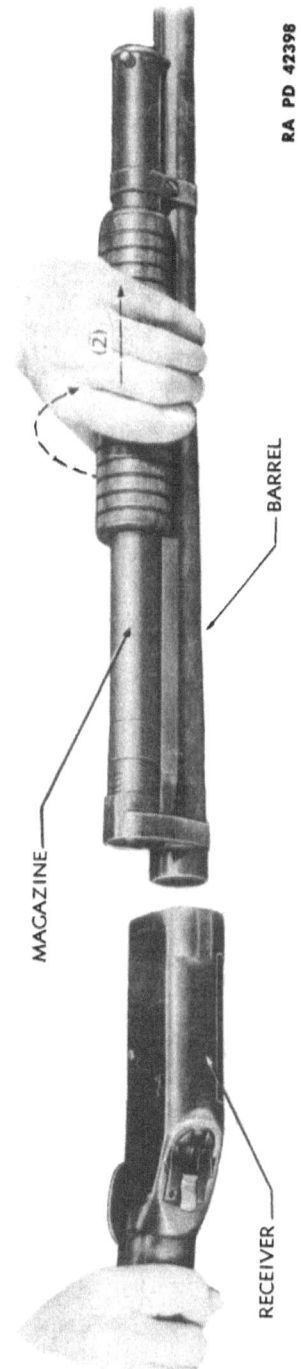

Figure 17 — Barrel and Magazine Disengaged from Receiver — Winchester Shotgun M97 (Take-down)

SHOTGUNS, ALL TYPES

with pin hole in guard bow and receiver and insert trigger pin. Drive pin through until flush.

(d) The butt stock can then be replaced by sliding on to tang of receiver and mating with groove in receiver. Then replace butt stock bolt and washer, and screw down tight by turning clockwise. The butt plate can then be replaced and screwed down snugly. Composition butt plates should be replaced carefully and not screwed down too tightly, to prevent cracking of plate and stripping of screw threads in butt stock.

(2) BREECH BOLT GROUP. The breech bolt group must be replaced in the receiver before the carrier group. Replace as follows:

(a) Insert the breech bolt group into the rear of the receiver with the locking lug forward and down, and push group all the way home.

(b) Slip action slide hook into left side of breech bolt locking lug, so that the boss on hook seats in counterbore in lug and hook lies in its guideway in left inner wall of the receiver, with point of hook down and to rear. Replace action slide hook screw from right side of breech bolt, and tighten. (Head of hook screw should protrude from right face of breech bolt at least $\frac{1}{32}$ inch to insure that it will cam down the shell guide, to clear the ejection opening in the receiver, when the breech bolt is retracted).

(c) Check action of breech bolt and hook by sliding breech bolt back and forth several times. Hook should move with breech bolt.

(3) CARRIER GROUP. The carrier group should not be replaced in the receiver until after the breech bolt has been replaced. Replace as follows:

(a) To replace the carrier group, first cock the hammer. Then, insert the carrier, hammer up and to the rear, into the receiver through the bottom opening, and keep moving to the rear into the receiver until the carrier seats itself properly in its recess in the rear wall of the receiver. Then, press firmly up on the carrier until it is completely home behind the locking lug of the breech bolt. In order to replace the carrier group properly, the breech bolt must be all the way forward and the shell guide stop screw must be removed from the right wall of the receiver.

(b) With the carrier in position, insert the carrier pin into its hole so that the groove on the pin is on the left side of the receiver and groove alined with the hole for the carrier pin stop screw, in the left top of the carrier body, to the left rear of the hammer. Replace stop screw.

(c) Replace the shell guide stop screw in its hole in the right wall of the receiver just ahead of the trigger pin. (This screw has a pin end, and its head when assembled, should be flush with the face of the receiver.

WINCHESTER SHOTGUN, 12-GAGE, M97

(4) BARREL, MAGAZINE, AND ACTION SLIDE GROUP (SOLID-FRAME GUN).

(a) With hammer forward in fired position, move breech bolt forward and lock carrier behind breech bolt by pressing upward on carrier until in position behind locking shoulder of bolt.

(b) With action slide assembled to magazine, hold magazine along under side of barrel, and insert rear end of slide bar into bar opening in left forward face of receiver, and push to rear so that it enters cam aperture in carrier.

(c) Mate threads in magazine with those in receiver, using care not to cross threads; screw magazine tightly into receiver by turning clockwise until stop screw hole in receiver and magazine aline; replace stop screw. An improvised tool can be used (par. 8 a (1) *(b)*) to tighten magazine.

(d) Replace bayonet attachment (par. 9 a (6)), then pull the action slide to the rear as far as it will go, and operate slide several times to test the operating mechanism and the locking of the breech bolt and slide.

(e) The barrel and magazine of the modified type of solid-frame gun (par. 4 c) are replaced as follows. The barrel is first replaced by inserting threaded end into receiver, with receiver extension facing to left, so that interrupted threads on barrel and in receiver are in a position to mate, and then by turning barrel ¼ turn counterclockwise, similarly to the take-down gun. The magazine with action slide assembled is then inserted through the receiver extension and into the receiver in a similar manner and turned ¼ turn counterclockwise. This gun can be identified by the receiver extension screwed to the rear end of the barrel, same as the take-down gun. Inspect bore of barrel for foreign matter before replacing.

(5) BARREL, MAGAZINE AND ACTION SLIDE GROUP (TAKE-DOWN GUN).

(a) Lock action as explained in 9 a (4) *(a)* above.

(b) Inspect bore for foreign matter. Then, push magazine and action slide forward on barrel as far as they will go, to clear rear face of receiver extension on barrel.

(c) Grasp barrel and magazine in left hand and receiver in right hand, and with magazine facing to the left, insert rear end of barrel into barrel aperture in forward face of receiver so that interrupted threads on barrel and receiver are in a position to mate. Push barrel into receiver as far as it will go and turn counterclockwise ¼ turn, thus mating interrupted threads on barrel with those in receiver. Turn barrel until stop screw in right forward face of receiver enters its aperture in right rear face of receiver extension on barrel and stops the barrel. The magazine aperture in receiver should now be in position to receive the magazine.

TM 9-285
9-10

SHOTGUNS, ALL TYPES

(d) With the barrel positioned, revolve the magazine until the interrupted threads on its rear end are in a vertical plane in line with barrel; then guide action slide bar into bar aperture in left forward face of receiver, and push magazine into receiver as far as it will go, so as to place interrupted threads on magazine and in receiver in position to mate. Then, revolve magazine counterclockwise ¼ turn, using the locking pin as a lever, until the depressed end of the locking pin is clear of the barrel on the left side. Then press locking pin flush with the magazine to lock it to the barrel, and hence prevent turning of the magazine. Locking pin should project from magazine on left side of barrel when magazine is locked.

(e) Pull the action slide handle to the rear as far as it will go, thus engaging the end of the slide bar with the action slide hook on the breech bolt, and the cam aperture in the carrier.

(f) Operate the slide handle several times to test functioning of the operating mechanism and the locking of the breech bolt and slide bar.

(6) BAYONET ATTACHMENT (fig. 5).

(a) Slide the bayonet attachment group onto the muzzle end of the barrel so that the sling swivel is to the rear and the sight up. Mate hole in rear of attachment lug with stud on magazine plug, and adjust attachment until the screw holes aline with the screw grooves in the underside of the barrel.

(b) Replace screws and screw in tightly. Check assembly for level seating and interference of hand guard with action slide handle when operated.

10. FIELD INSPECTION.

a. With gun completely assembled, test mechanism for proper functioning. Fired shells may often be used for testing, where dummy shells are not available, by turning in the uncrimped end so that the length of the shell approximates that of a live shell. Use of live shells for testing is prohibited.

CAUTION: Be sure gun is fully unloaded before inspection.

b. **Operate the Gun as Follows:**

(1) With the breech bolt locked and the hammer at full cock, press in the action slide lock release plunger pin, located on the right side of the receiver. Push action slide handle forward slightly and then pull smartly and fully to the rear, and then push smartly and fully forward. Reciprocate action slide thus several times to test smoothness of action.

(2) With breech bolt locked, let hammer down to half cock; press release plunger pin and attempt to retract action slide handle as above. The action slide handle should not retract.

WINCHESTER SHOTGUN, 12-GAGE, M97

(3) Place hammer at full cock; pull trigger to allow hammer to move fully forward to the fired position and attempt to retract action slide handle. The action slide handle should retract.

(4) Retract action slide handle as in (1) above; then release plunger pin and push handle smartly and fully forward to lock the breech bolt. Then attempt to retract action slide handle. It should not be possible to do so with hammer at full cock until action slide lock is disengaged as in (1) above.

(5) With breech bolt locked and hammer at full cock, pull trigger to test firing mechanism. Then, place hammer at half cock and attempt to pull the trigger. It should not be possible to pull the trigger. When hammer is at half cock the firing mechanism is in the safe position.

(6) With hammer at full cock, close but do not lock the breech bolt. Then attempt to fire the gun. The gun should not fire until the breech bolt is locked.

(7) Place two or more dummy or fired shells in the magazine and work through the action to test gun for feeding, loading, extraction and ejection of shells. The second shell should not leave the magazine until after the carrier has dropped as the first shell is being ejected.

NOTE: Fired shells will not work through the action as easily as live or dummy shells, as they are somewhat deformed through being fired. Therefore allowance should be made for friction and smoothness of action in positioning the shell.

c. When gun does not operate and function smoothly and properly when tested as above, damaged or improperly assembled parts are indicated as follows:

(1) ACTION SLIDE STICKS. May be due to bent slide bar, burs on bar cam lug, foreign matter in camming aperture in carrier, or burs on breech bolt guideways or guides.

(2) BREECH BOLT DOES NOT LOCK. May be due to foreign matter on face of bolt or in extractor cuts in barrel, behind locking shoulder of bolt or on carrier, or worn or burred action slide bar cam lug.

(3) HAMMER DOES NOT COCK PROPERLY OR SLIPS. May be due to burs or foreign matter in sear notches, burred sear nose, weak or broken sear spring, or foreign matter between trigger and sear.

(4) FIRING PIN DOES NOT RETRACT IN BREECH BOLT. May be due to broken or missing lock spring, foreign matter in breech bolt, or broken parts.

(5) ACTION SLIDE LOCK DOES NOT FUNCTION. May be due to foreign matter under lock, broken lock or lock spring, burs on lock or action slide bar. Refer to CAUTION at end of paragraph 3.

SHOTGUNS, ALL TYPES

(6) SHELL IS NOT EXTRACTED OR EJECTED. May be due to broken or worn extractors or ejector, or broken springs.

(7) TWO SHELLS FED INTO RECEIVER AT ONCE. May be due to broken shell stop or springs, or foreign matter under stops.

(8) SHELLS STICK IN MAGAZINE. May be due to corroded or bent follower, dented tube, broken or kinked spring, or foreign matter or corrosion in tube.

d. Inspect barrel and test trigger pull (par. 3 n).

e. In addition to inspection of the gun for operation and functioning, the gun should be inspected generally for condition, and defects noted. Attention should be directed to such defects as cracked wooden parts, cracked or deformed metal parts, dented magazine tube, loose screws, loose or binding parts or assemblies, loose barrel or magazine, loose stock or butt plate, rust, dents, burs, or excessive wear of parts. If defects are such that early malfunction of the gun is indicated, the gun should be turned over to ordnance personnel for inspection and correction.

f. Where defects and malfunctions cannot be remedied by cleaning, lubrication, and simple adjustments of assembly, which lie within the scope of using troops, the gun should be turned over to ordnance personnel for a thorough inspection, correction, and/or repair.

g. Removal of burs on working parts, trigger adjustments, and like corrections should not be attempted by using troops as stoning of parts must be exacting, the angle of the faces concerned must not be changed, and volume of metal must not be materially reduced.

h. A loose barrel, which shakes when assembled, may be due to improper assembly caused by not inserting barrel (take-down type) far enough into receiver before mating interrupted threads of barrel and receiver, or it may be due to worn parts necessitating adjustment or replacement. If due to worn parts, gun should be turned over to ordnance personnel for correction or repair.

i. If shell appears unnecessarily loose in chamber with breech bolt locked, the gun should be turned over to ordnance personnel to be checked for headspace.

j. Adjustment and maintenance of the gun in the case of using troops is limited to the removal and replacement of the parts and groups of parts as outlined in paragraphs 8 and 9, together with cleaning and lubrication, and such adjustments as are necessary in assembling the gun as outlined.

11. CLEANING AND LUBRICATION.

a. **Cleaning the Bore.**

(1) Barrel and magazine group of take-down type gun should be

WINCHESTER SHOTGUN, 12-GAGE, M97

removed from the receiver when cleaning the bore. Bore of solid-frame gun may be cleaned from the breech end of the receiver when the breech bolt and carrier have been removed, or from the muzzle end. Ordinarily the bore should be cleaned from the muzzle end, and a rag stuffed into the receiver to protect the action. If bore of solid-frame gun is badly rusted or corroded, the gun should be disassembled and the barrel thoroughly cleaned from the breech end.

(2) The bore should be thoroughly cleaned with **CLEANER**, rifle bore, or soap and water solution applied to a cloth patch assembled to the cleaning rod; then thoroughly dried and lightly oiled with **OIL**, lubricating, preservative, light.

CAUTION: All oil should be removed from bore and chamber before firing the gun.

b. **Cleaning Parts Other Than the Bore.**

(1) Groups should be removed and parts and assemblies thoroughly cleaned, oiled, and lubricated at regular intervals or when gun has been exposed to extreme conditions such as moisture, dust storms, or the like. Pointed sticks may be used to clean crevices and like inaccessible points.

(2) Cleaning is best accomplished by wiping parts with a rag slightly moistened with **CLEANER**, rifle bore, or light oil to loosen burnt powder and like foreign matter. Then wipe clean with clean dry rag, and oil lightly, using rag lightly moistened with **OIL**, lubricating, preservative, light.

(3) Special attention should be paid to cam and guide grooves, spring seats, firing pin aperture, and like apertures where foreign matter may become lodged and caked and prevent proper functioning of the mechanism.

(4) Wood parts of stock and action slide handle may be cleaned by wiping with slightly oily rag and then polishing with clean dry rag. When wooden parts have a rubbed down oil finish without varnish, a light application of **OIL**, linseed, raw, applied with a rag and well rubbed into the wood with the heel of the hand, will help to preserve the wood. Care should be observed to prevent linseed oil from getting on metal parts or into mechanism of gun as it will become gummy when dry.

(5) Inside of magazine tube should occasionally be wiped out. This can be accomplished without disassembling the magazine by assembling a cleaning patch to the cleaning rod, moistening *lightly* with light lubricating oil and inserting in rear end of magazine against the follower. Push patch back and forth in tube a few times, using care that patch does not tear off or become caught in magazine. Excess oil should be

TM 9-285

SHOTGUNS, ALL TYPES

avoided as it will penetrate shells and/or get into chamber of barrel, thus causing excessive pressure at this point.

c. **Cleaning Gun in Garrison or in Camp.**

(1) If gun has been fired considerably it should be thoroughly cleaned as prescribed above.

(2) If, however, only a few shots have been fired and the gun has not been subject to extreme conditions such as dust or rain, it will usually suffice to clean bore and then wipe receiver, barrel and accessible metal parts, outside and in with a rag lightly saturated with OIL, lubricating, preservative, light, without removing the groups from the receiver.

(3) If the gun has not been fired and is otherwise clean and dry, and it is expected to be used the following day it should be wiped thoroughly as above with a rag very lightly moistened with oil. Oil may then be removed from bore and chamber and the outside surfaces of the gun with a dry rag before use. If gun is to be used for guard duty where immediate firing is not anticipated, a *light* film of oil may be allowed to remain in the bore. If, however, the gun is not to be used for a few days, it should be well oiled, and inspected daily and oil film renewed when necessary. For cleaning and oiling preparatory to long or short term storage, and for cleaning as received from storage, refer to paragraphs 77 and 78.

d. **Rust and Corrosion.**

(1) Gun should be kept free of rust or corrosion at all times. Light rust may usually be removed with an oily rag or one moistened with **CLEANER**, rifle bore. If this method does not suffice, **CLOTH**, crocus, or **WOOL**, steel, fine, may be used. Care should be exercised to prevent undue scratching of surfaces.

(2) Heavy rusting or leading of the bore of the barrel may at times occur. Rust usually appears in dark irregular patches, while leading shows in dull gray streaks. Leading is due to a small quantity of lead from the shot pellets adhering to rough spots on the inner surface of the barrel, and seldom occurs when chilled shot is used. Rust and leading in the bore may be removed with a wad of WOOL, steel, fine, on the end of a cleaning rod. The wool should be pushed the full length of the bore each time and on the bore line, and not turned, or the bore scrubbed. This method prevents scratching the barrel which hastens fouling.

e. **Lubrication.**

(1) In addition to light oiling for preservation, gun should be lightly lubricated, using OIL, lubricating, preservative, light. Lubrication should be kept to a minimum as too much oil attracts foreign matter and burnt powder which will become caked and cause malfunction or undue wear

WINCHESTER SHOTGUN, 12-GAGE, M97

of the mechanism. Lubricating is best accomplished by using an oiler with a small nose, or a dropper, by which the direction and amount of oil used can be controlled.

(2) POINTS TO LUBRICATE ARE:

(a) Action slide bar opening in forward end of receiver.

(b) Action slide bar cam lug.

(c) Action slide lock.

(d) Carrier pin.

(e) Breech bolt guides.

(f) Outer surface of magazine tube where action slide bears.

(g) Action slide hook screw.

(h) Hammer pin.

(i) Action slide lock release, in carrier.

(3) In very cold climates, oiling and lubrication should be reduced to a minimum. Only surfaces showing signs of wear should be *lightly* oiled. Refer to "Special Maintenance," section XI.

TM 9-285

SHOTGUNS, ALL TYPES

Section III

WINCHESTER SHOTGUN, 12-GAGE, M12

	Paragraph
Description	12
Data	13
Operation	14
Functioning	15
Removal of groups	16
Replacement of groups	17
Field inspection	18
Cleaning and lubrication	19

12. DESCRIPTION.

a. Identification marks on this gun are generally to be found as follows:

(1) Name of maker, gage, model, and barrel boring are stamped on the top of the barrel near the breech end.

(2) The serial number of the gun is stamped on the lower face of the barrel near the breech end, and on the lower face of the forward end of the receiver.

b. This gun (figs. 18 and 19) is a manually operated repeating shotgun of the slide action, hammerless, solid-frame and take-down type. The take-down gun is so constructed that the barrel and magazine together with the action slide group can easily be removed as a unit by disengaging interrupted threads on rear end of magazine and barrel from like interrupted threads in the receiver, by which the barrel and magazine are locked to the receiver. This construction facilitates cleaning and transportation of the gun. The construction of barrel magazine and action slide group is similar to, but *not* interchangeable with the Winchester M97 take-down type of gun described in section II. The receiver and operating mechanism, however, differ widely in design. The solid frame type of gun, when issued, will be constructed with barrel and magazine similar to the modified solid-frame Winchester M97 gun (par. 4 c).

c. This gun is furnished in various grades having barrels of different lengths and degrees of boring and other modifications of design. Basically, however, the mechanism of all guns of this make and model are the same. For convenience herein the guns will be classified as three types: riot, sporting skeet, and sporting trap; although variations of these types may occur.

WINCHESTER SHOTGUN, 12-GAGE, M12

(1) The riot-type gun is usually furnished with a 20-inch plain barrel, bored cylinder, and may be equipped with a bayonet attachment and hand guard attached to the muzzle end of the barrel, and a leather sling attached to a sling swivel on the bayonet attachment and the stock. If equipped with bayonet attachment it is solid-frame. Refer to paragraph 3 i.

(2) The sporting skeet-type gun is usually furnished with a 26-inch plain or ribbed barrel, bored improved cylinder, and is without bayonet attachment or sling.

(3) The sporting trap-type gun (figs. 18 and 19) is similar to the skeet-type gun but is furnished with a 30-inch barrel, bored full choke, and is without bayonet attachment or sling.

d. **General Description** (figs. 20, 21, 26 and 27).

(1) The stock of the gun is bolted to the rear end of the receiver and the barrel and magazine locked to the forward end of the receiver as already explained in paragraph b above. The action slide is mounted and operates on the magazine. The rear end of the slide or bar passes through the forward end of the receiver, and engages with, reciprocates, and cam-operates the breech bolt, which in turn operates the carrier and cams back and cocks the hammer.

(2) The receiver contains the operating mechanism; and to its lower rear end is attached the trigger guard to which is mounted the firing mechanism, carrier and action slide lock. The receiver is open at the bottom to permit loading, and the right side for ejection of the fired shell cases. The ejector is seated in the inner left wall of the receiver.

(3) The breech bolt contains the extractors, the firing pin, and firing pin retractor by which the firing pin is cammed back from the face of the bolt and blocked from being driven forward by the hammer until the bolt is locked in position.

(4) The trigger guard contains the hammer, trigger, trigger (safety) lock, carrier, and action slide lock mechanism together with their springs and components.

(5) The magazine which is of the tubular type is positioned beneath the barrel and has a capacity of five shells loaded end to end. The shells are pressed together and fed into the receiver by the force of the magazine spring acting upon the follower.

(6) The shell cut-off is pivoted in the left inner wall of the receiver, just to the rear of the magazine opening and acts to hold the shells in the magazine against the pressure of the magazine spring. The cut-off is operated by the action slide bar to release and block the shells at the proper time, thus allowing but one shell to enter the receiver at a time. The carrier assumes the function of the cut-off while the latter is cammed back from the blocking position by the action slide bar.

SHOTGUNS, ALL TYPES

Figure 18 — Right Side View — Sporting Trap Type — Winchester Shotgun M12

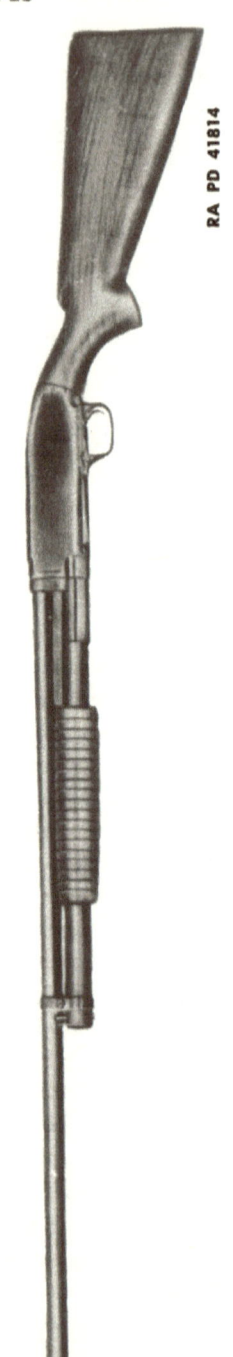

Figure 19 — Left Side View — Sporting Trap Type — Winchester Shotgun M12

WINCHESTER SHOTGUN, 12-GAGE, M12

(7) The action slide lock (figs. 24 and 25) is pivoted in the left side of the trigger guard, and engages with the action slide bar to block its rearward movement after the breech bolt is locked in position by the action slide bar, thereby preventing premature unlocking of the breech bolt. The lock is disengaged either by the descending hammer when the gun is fired, or by manual upward pressure on the rear end of the action slide lock projecting through the guard to the left rear of the trigger. A hook on the lock engaging with a lug on the side of the hammer locks the hammer in the extreme rearward position and holds it from engagement with the sear until the breech bolt is locked in place. This prevents premature firing of the gun.

NOTE: The sear is an integral part of the trigger in this gun.

(8) The trigger (safety) lock is positioned in the forward end of the trigger guard bow and acts to block or clear the trigger, thus preventing or allowing retraction, to fire the gun.

13. DATA.

Gage of bore	12
Boring of barrel—riot type	Cylinder
Boring of barrel—sporting skeet type	Improved cylinder
Boring of barrel—sporting trap type	Full choke
Type of action	Slide
Type of firing mechanism	Hammerless
Type of magazine	Tubular
Capacity of magazine	5 shells
Length of barrel—riot type	20 in.
Length of barrel—sporting skeet type	26 in.
Length of barrel—sporting trap type	30 in.
Length of stock and receiver (approx.)	20 in.
Length of assembled gun—riot type (approx.)	40 in.
Length of assembled gun—sporting skeet type (approx.)	46 in.
Length of assembled gun—sporting trap type (approx.)	50 in.
Weight of assembled gun—riot type without bayonet or attachment (approx.)	6½ lb
Weight of assembled gun—sporting skeet type (approx.)	7 lb
Weight of assembled gun—sporting trap type (approx.)	7⅜ lb
Weight of bayonet M1917 (approx.)	1⅛ lb

14. OPERATION.

a. The gun is operated by moving the action slide handle smartly and fully, backward and forward. This action unlocks the breach bolt, extracts and ejects the fired shell, cocks the hammer, transfers a live shell from the magazine to the chamber of the barrel and relocks the breech bolt behind the shell.

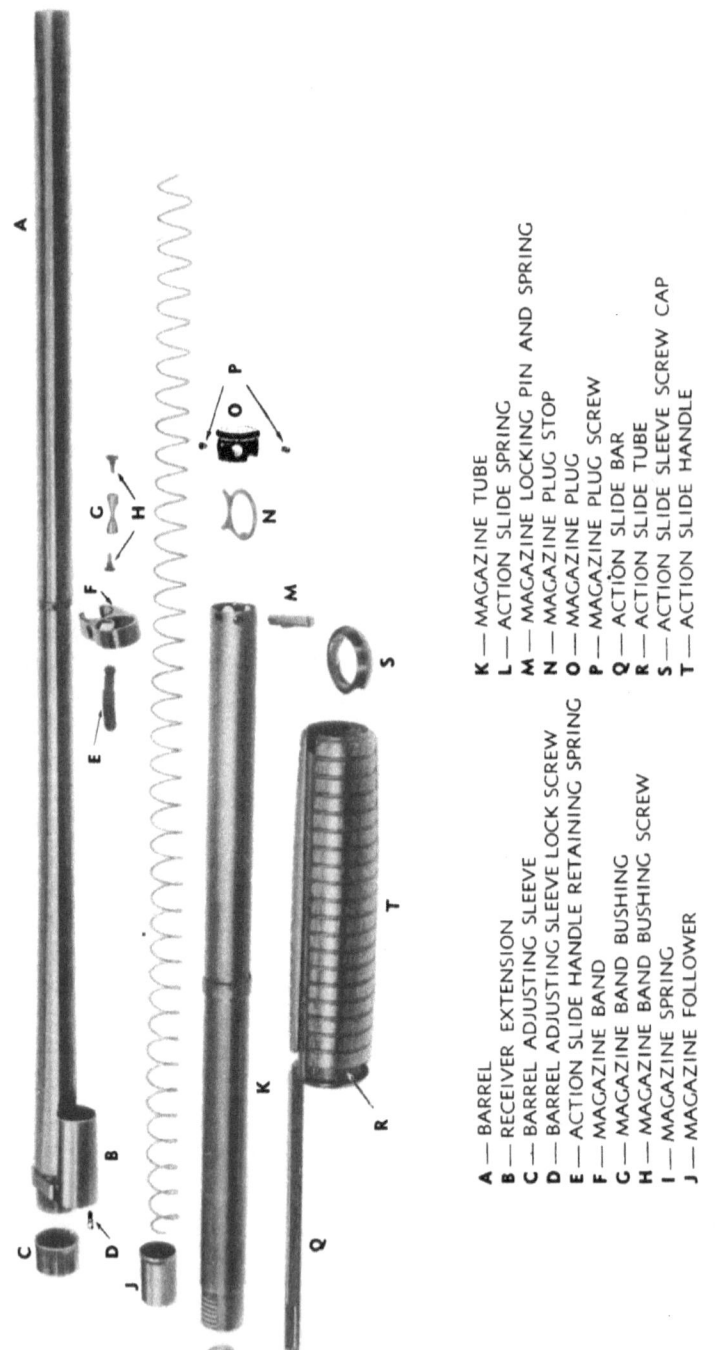

Figure 20—Barrel, Magazine and Action Slide Groups—Disassembled View—Winchester Shotgun M12 (Take-down)

TM 9-285
14

WINCHESTER SHOTGUN, 12-GAGE, M12

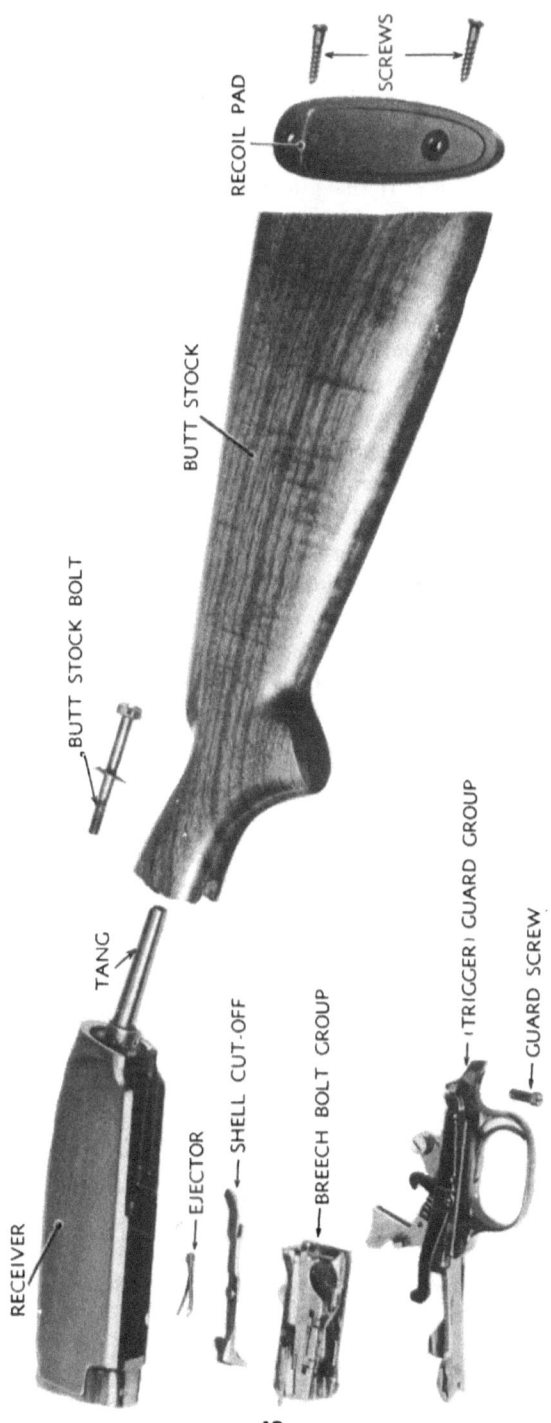

Figure 21—Receiver and Stock Groups—Disassembled View—Winchester Shotgun M12 (Take-down)

SHOTGUNS, ALL TYPES

CAUTION: During operation, the muzzle of the gun should always be pointed at a safe spot.

b. Before the action slide can be retracted, however, the action slide lock must be disengaged from the action slide bar (fig. 28). To accomplish this, the action slide handle must be moved slightly forward to allow the lock to disengage. If the gun has been fired, and consequently the hammer is forward, the only movement necessary is to move the action slide handle forward slightly, and then reciprocate it as above, as the descending hammer has already partly disengaged the lock. If, however, the hammer is in the cocked position, it is necessary to press up the rear end of the action slide lock which is visible in the rear lower face of the receiver just to the left of the trigger before moving the action slide handle.

c. When the gun is being fired as a repeater, the recoil of the gun performs the preliminary forward movement of the action slide handle, as the gun recoils away from the handle which is held by the operator.

CAUTION: During these operations, the finger should remain outside the trigger guard. Reciprocation of the action slide handle should be smart and full to insure extraction of the shell, cocking of the hammer and complete locking of the breech bolt and engagement of the action slide lock. Slamming of the mechanism, however, should be avoided. When the gun is being fired as a repeater, all pressure should be removed from the trigger while operating.

d. With the gun loaded, locked, and cocked, the only operation necessary to fire the gun is the retraction of the trigger.

e. The trigger (safety) lock (fig. 22), positioned in the forward part of the trigger guard bow, operates laterally. When pushed to the left, so that the red band is visible, the trigger is free to be pulled and the gun fired. When pushed to the right, the trigger is blocked and cannot be pulled nor the gun fired. With a shell in the chamber, it is best to lock the trigger by pushing the lock to the right, unless the gun is to be fired immediately.

f. **To Load and Unload the Magazine.**

(1) To load the magazine (fig. 23), press a shell, nose first, into the rear of the magazine against the magazine follower, until it slips in front of and is retained by the carrier. Load another shell in the same way until five in all are loaded. Loading should be done with breech bolt locked.

(2) To unload magazine, press up the carrier, with breech bolt locked, and allow the shells to slip out one by one.

g. **To Load and Unload the Gun.**

(1) To load a shell from the magazine into the chamber, push trigger

WINCHESTER SHOTGUN, 12-GAGE, M12

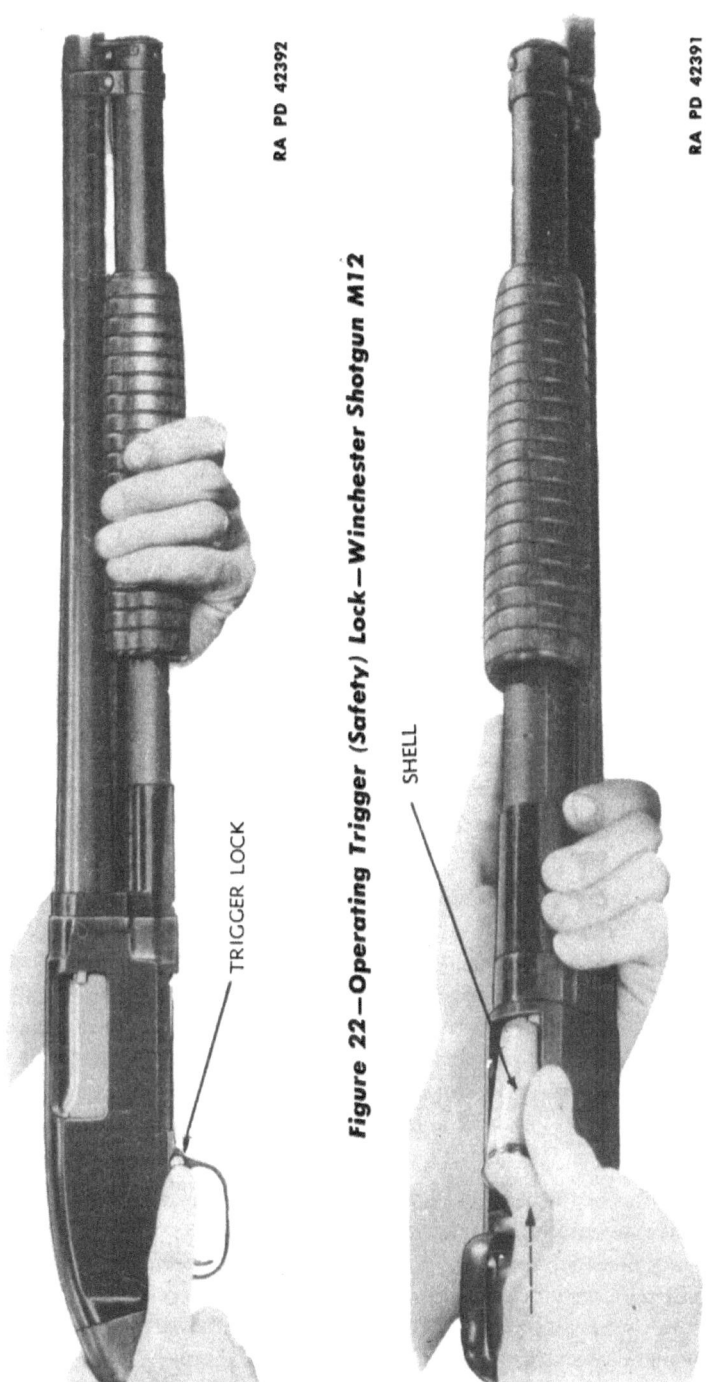

Figure 22—Operating Trigger (Safety) Lock—Winchester Shotgun M12

Figure 23—Loading Magazine—Winchester Shotgun M12

SHOTGUNS, ALL TYPES

lock to the right, push up the rear end of the action slide lock, move action slide handle slightly forward, and then reciprocate. Another shell may then be loaded into the magazine. Allow trigger (safety) lock to remain in safe position to lock trigger unless gun is to be fired immediately.

(2) To unload the gun, slide trigger (safety) lock to the right as above and unload the magazine (par. 14 f (2)). Then disengage the action slide lock and retract action slide handle to extract and eject the shell in the chamber. Inspect magazine and chamber to make sure gun is fully unloaded.

h. To Load Chamber Only. With magazine empty and trigger (safety) lock pushed to right, pull back action slide handle to retract breech bolt all the way. Then place shell in chamber of barrel through ejection opening of receiver and lock the breech bolt by pushing action slide handle all the way forward.

15. FUNCTIONING.

a. As already explained the functioning of the operating mechanism is accomplished by the reciprocation of the action slide handle. A cam lug on the rear end of the action slide bar engages in an irregular camming aperture in the left side of the breech bolt, and as the slide is moved rearward the rear end of the breech bolt is cammed down from in front of the locking shoulder in the top of the receiver and the bolt moved to the rear. As the action slide is moved forward, the breech bolt is moved forward and its rear end cammed up in front of the locking shoulder at the end of the forward movement.

b. As the breech bolt moves to the rear, it cams back the hammer which is caught and held by a hook on the action slide lock engaging with a lug on the side of the hammer. The hammer is thus held from engagement with the sear (fig. 24). As the breech bolt reaches the forward position, the action slide lock, which has been held depressed by the action slide bar, is released and springs up behind the action slide bar to block its rearward movement (fig. 25). At the same time, the lock releases the hammer which moves forward and is caught and held by the sear.

c. As the breech bolt starts forward, it cams the carrier up and down rapidly. This movement lifts the shell, resting on the carrier, in line with the bore and the shell is pushed forward by the breech bolt into the chamber of the barrel. The carrier has immediately returned to the lower position and is in readiness to receive another shell. As the breech bolt closes, a camming shoulder in the action slide bar, acting upon the shell cut-off, cams it downward, thus releasing a shell from the magazine which

WINCHESTER SHOTGUN, 12-GAGE, M12

is pushed into the receiver by the force of the magazine spring and held by the carrier.

d. The fired shell is extracted from the chamber of the barrel by the extractors in the forward end of the breech bolt, as the bolt moves to the rear, and knocked out of the receiver through the ejection opening in the right side, as the base of the shell strikes the ejector positioned in the inner left wall of the receiver.

16. REMOVAL OF GROUPS (figs. 20, 21, 26 and 27).

a. Groups and parts should be removed and replaced in the order given below. Groups and parts when removed should be placed on a clean, flat surface and care observed to prevent loss of screws and small parts. Remove as follows:

(1) BAYONET ATTACHMENT, BARREL, MAGAZINE AND ACTION SLIDE GROUPS (SOLID-FRAME GUN). When necessary to remove these groups from the solid-frame gun, it may be accomplished in a manner similar to that prescribed for the Winchester M97 gun of the same type (par. 8 a 1(a)(c)).

(2) BARREL, MAGAZINE AND ACTION SLIDE GROUP (TAKE-DOWN GUN).

(a) Lock the breech bolt by pushing the action slide handle all the way forward.

(b) Unlock the magazine by pressing the locking pin through the plug until it will clear the barrel when the magazine is turned (fig. 29).

(c) Turn the magazine ¼ turn clockwise, using the pin as a lever, to disengage the interrupted threads on the rear end of the magazine from those in the receiver. Then, pull the magazine forward out of engagement with the receiver.

(d) Push the action slide forward until the rear end of the bar is free of the receiver. Then, grasp barrel in left hand and receiver in right hand and turn the barrel ¼ turn clockwise, to disengage the interrupted threads on the rear of the barrel from those in the receiver, and withdraw the group forward from the receiver (figs. 30 and 31).

(3) TRIGGER GUARD GROUP.

(a) The guard group may be removed from the receiver without removing the butt stock.

(b) To remove the guard group, unscrew guard screw from rear end of guard with receiver bottom side up and pull guard, with carrier attached, upward and to the rear out of the receiver.

TM 9-285
16

SHOTGUNS, ALL TYPES

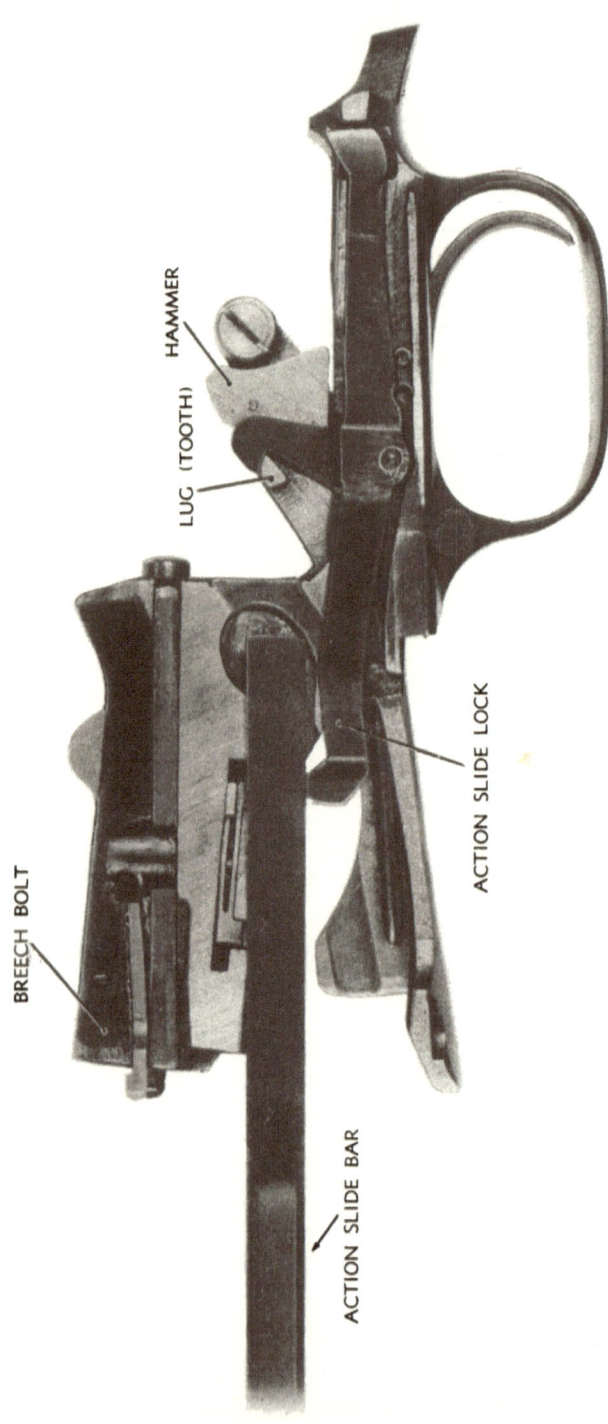

Figure 24—Breech Bolt (Trigger) Guard and Action Slide Bar Groups—Action Slide Lock Disengaged—Hammer Locked—Winchester Shotgun M12

TM 9-285

WINCHESTER SHOTGUN, 12-GAGE, M12

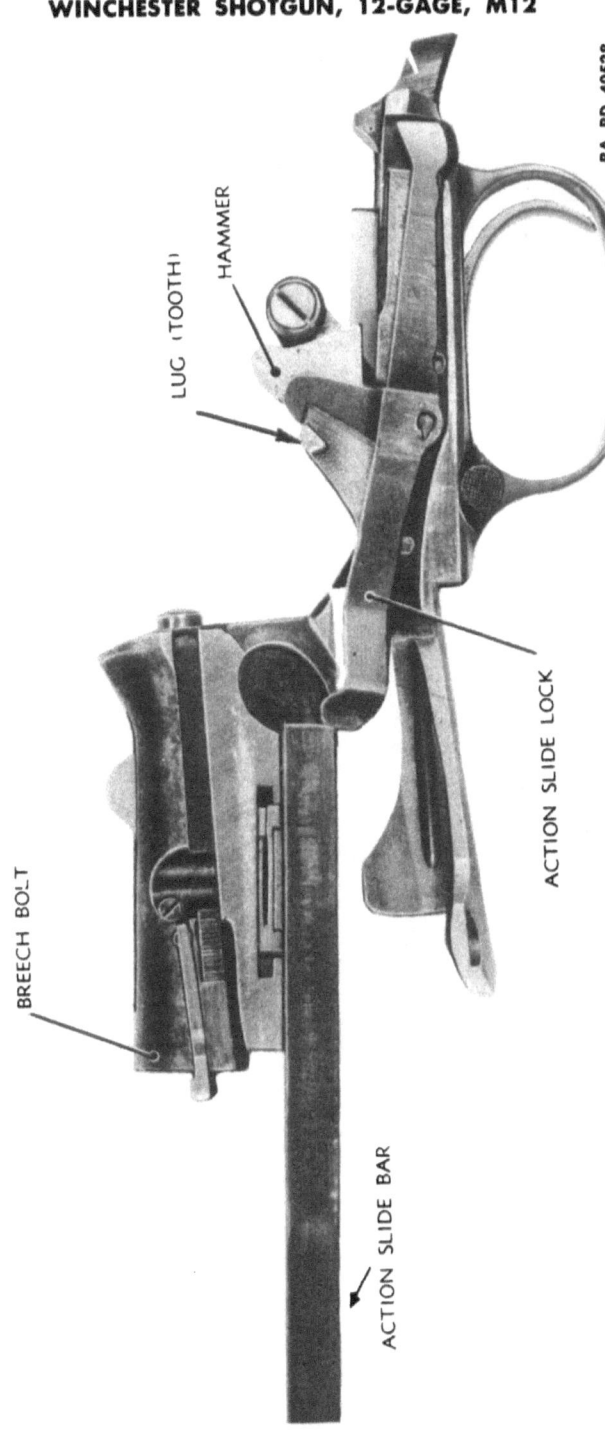

Figure 25—Breech Bolt (Trigger) Guard and Action Slide Bar Groups—Action Slide Lock Engaged—Hammer Unlocked—Winchester Shotgun M12

TM 9-285

SHOTGUNS, ALL TYPES

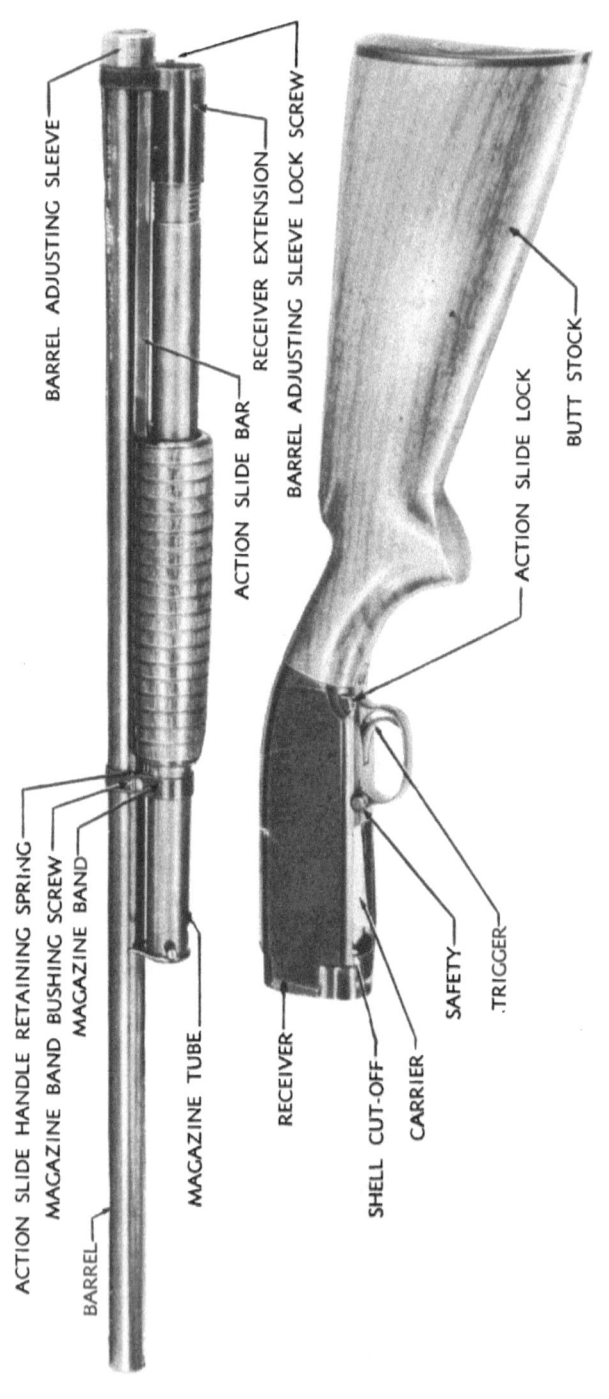

Figure 26 — Gun Taken Down — Left Side View — Showing Location of Parts — Winchester Shotgun M12 (Take-down)

WINCHESTER SHOTGUN, 12-GAGE, M12

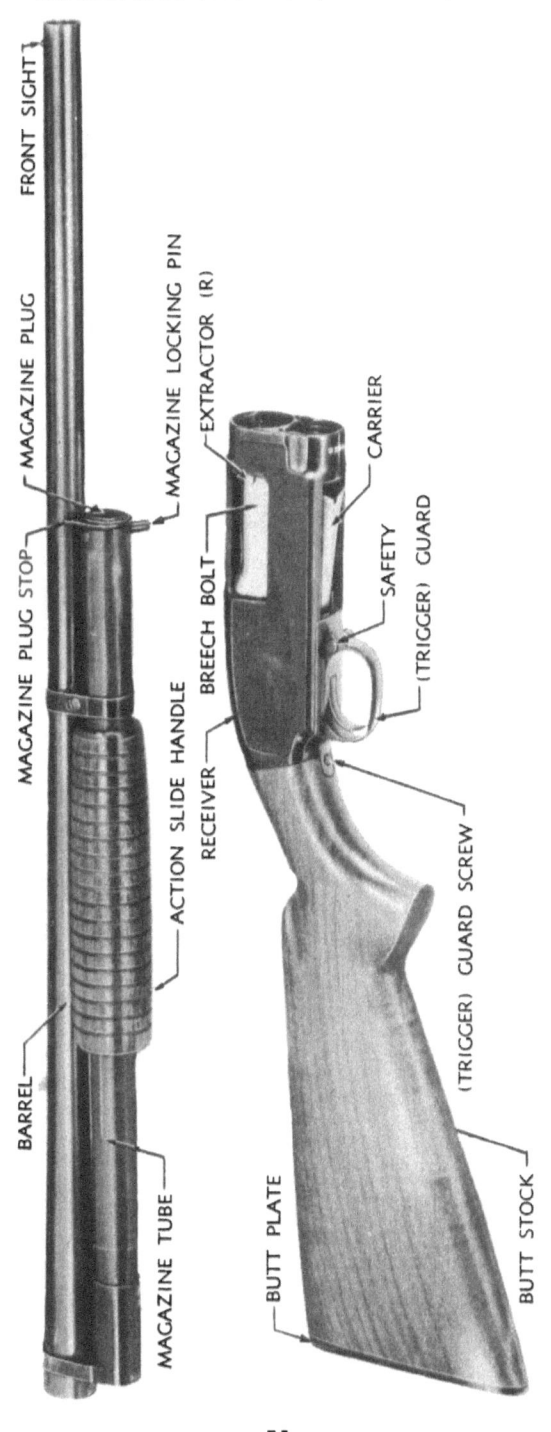

Figure 27 — Gun Taken Down — Right Side View — Showing Location of Parts — Winchester Shotgun M12 (Take-down)

TM 9-285
SHOTGUNS, ALL TYPES

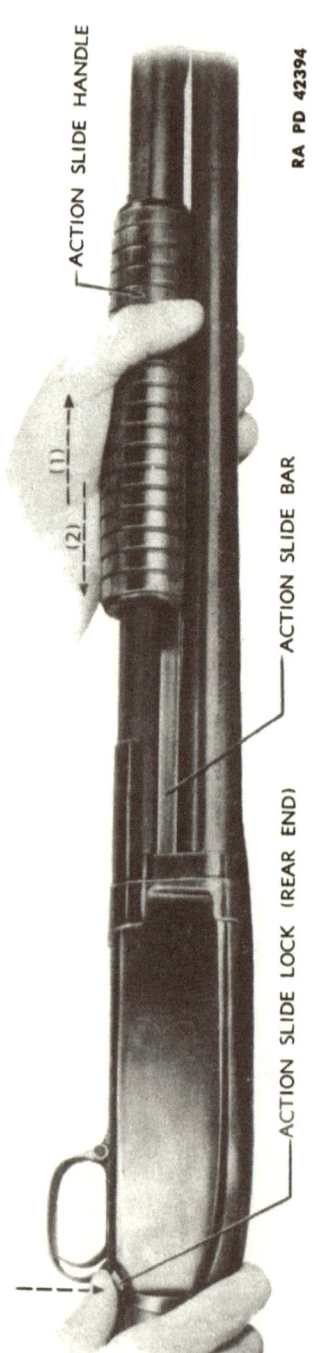

Figure 28—Disengaging Action Slide Lock from Action Slide Bar—Winchester Shotgun M12

Figure 29—Unlocking Magazine—Winchester Shotgun M12 (Take-down)

TM 9-285
16

WINCHESTER SHOTGUN, 12-GAGE, M12

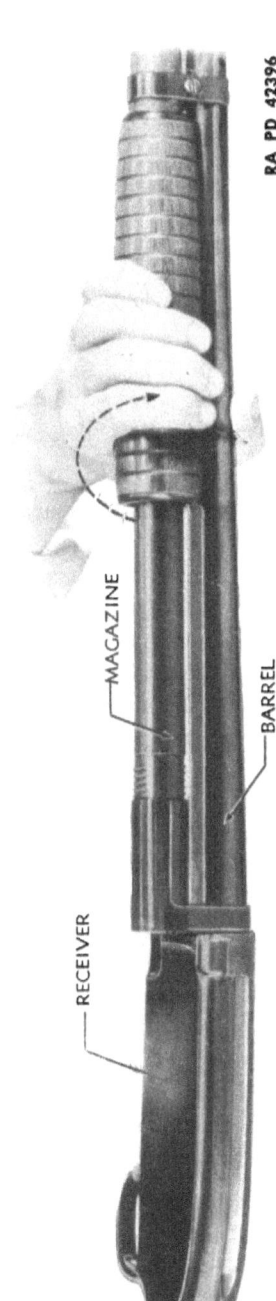

Figure 30 — Unlocking Barrel from Receiver — Magazine Already Disengaged — Winchester Shotgun M12 (Take-down)

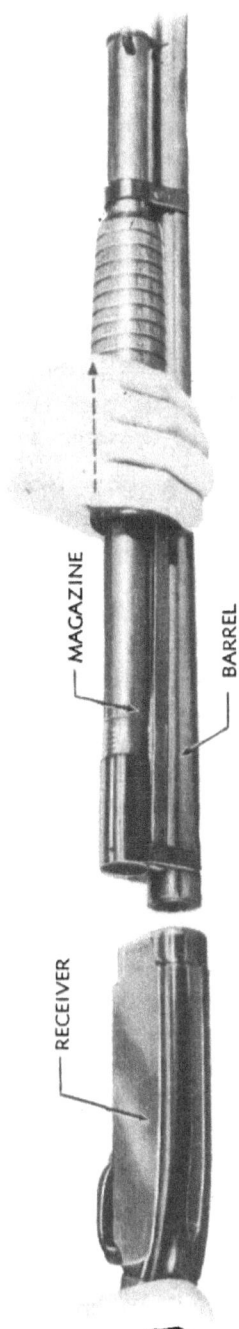

Figure 31 — Barrel and Magazine Disengaged from Receiver — Winchester Shotgun M12 (Take-down)

SHOTGUNS, ALL TYPES

(4) BREECH BOLT GROUP.

(a) With the guard group removed from the receiver, and the breech bolt in the forward position, pry out the ejector from the left inner wall of the receiver.

(b) Pry out the shell cut-off from the lower left inner wall of the receiver, if it has not already fallen out.

(c) Reach into the receiver with a screwdriver or similar tool, and depress the rear end of the breech bolt retaining lever, positioned in the left face of the breech bolt. Then, slide the breech bolt to the rear and lift from the receiver.

17. REPLACEMENT OF GROUPS.

a. Groups and parts should be thoroughly cleaned, lightly oiled, and lubricated, if necessary, before replacing. Replace as follows:

(1) BREECH BOLT GROUP.

(a) The breech bolt must be replaced before the guard group, the ejector, and shell cut-off.

(b) With the receiver bottom side up, drop the breech bolt, rounded face down and extractors forward, into the rear of the receiver. Push forward to the locked position and then, with a screwdriver or similar tool, depress the forward end of the breech bolt retaining lever positioned in the left face of the breech bolt. The lever is held in position by friction and should work stiffly. Test before replacing bolt. (The rear end of the bolt must be fully seated before the rear end of the lever will enter its groove in receiver. Do not force).

(c) With the breech bolt in the locked position, insert the ejector into its aperture in the left inner wall of the receiver to the rear of the breech bolt so that disk-shaped head of ejector is to the rear and the spring towards the receiver. Press down into aperture until head is flush with inner face of receiver.

(d) Insert shell cut-off into its aperture in lower left, inner wall of the receiver just to the rear of the magazine opening so that the enlarged curved end is forward and fitted into mating groove in the receiver to the left of magazine opening, and the cylindrical stud fits into its pivot hole in the receiver. The cut-off should lie flush in its groove when assembled. The cut-off is liable to fall out, until the trigger guard is in place.

(2) TRIGGER GUARD GROUP.

(a) The guard group should not be replaced in the receiver until the breech bolt has been replaced and shifted to the forward locked position.

(b) With the hammer cocked, grasp the guard by the bow and insert the forward end of the guard downward and forward at a slight angle into the receiver. Mate the forward end of the guard with the receiver

WINCHESTER SHOTGUN, 12-GAGE, M12

and then press down the rear end until it slides into the receiver, and the screw holes in guard and receiver aline. When properly mated, guard should position easily. Do not force.

(c) Replace guard screw and turn down snugly.

(3) BAYONET ATTACHMENT, BARREL, MAGAZINE AND ACTION SLIDE GROUPS (SOLID-FRAME GUN). If these groups have been removed from the gun, they may be replaced in a manner similar to that prescribed for the Winchester M97 gun of the same type (par. 9 a (4) (e) and (6)).

(4) BARREL, MAGAZINE AND ACTION SLIDE GROUPS (TAKE-DOWN GUN).

(a) Move breech bolt to the forward locked position, by pushing forward and up.

(b) Inspect bore for foreign matter. Then push magazine and action slide forward on barrel as far as they will go, to clear rear face of receiver extension on barrel.

(c) Grasp barrel in left hand and receiver in right hand, and with magazine facing to the left, insert threaded end of barrel into barrel aperture in forward face of receiver so that interrupted threads on barrel and receiver are in a position to mate. Push barrel into receiver as far as it will go and turn counterclockwise ¼ turn, thus mating interrupted threads on barrel with those in receiver. Turn barrel until barrel adjusting bushing lock screw, in right rear face of receiver extension on barrel, enters its aperture in right forward face of receiver and stops the barrel. The magazine aperture in receiver should now be in position to receive the magazine.

(d) With the barrel positioned, revolve the magazine until the interrupted threads on its rear end are in a vertical plane in line with barrel; then guide action slide bar into bar aperture in left forward face of receiver, and push magazine into receiver as far as it will go so as to place interrupted threads on magazine and in receiver in position to mate. Then, revolve magazine counterclockwise, using the locking pin as a lever, until the depressed end of the locking pin is clear of the barrel. Then, press locking pin flush with the magazine to lock it to the barrel, and hence prevent turning of the magazine. Locking pin should project from the magazine on left side of barrel when magazine is locked.

(e) Pull the action slide handle to the rear as far as it will go, thus engaging it with the camway in the breech bolt.

(f) Operate the slide several times to test functioning of the operating mechanism and the locking of the breech bolt and slide.

18. FIELD INSPECTION.

a. With the gun completely assembled, test the mechanism for proper functioning. Fired shells may often be used for testing, where dummy

SHOTGUNS, ALL TYPES

shells are not available, by turning in the uncrimped end so that the length of the shell approximates that of the live shell. Use of live shells for testing is prohibited.

CAUTION: Be sure gun is fully unloaded before inspection.

b. **Operate the Gun as Follows:**

(1) With the breech bolt locked and the hammer cocked, press upward on the rear end of the action slide lock, showing at the left rear of the trigger guard bow. Push action slide handle forward slightly, then pull smartly and fully to the rear, and then push smartly and fully forward. Reciprocate action slide handle thus several times to test smoothness of action.

(2) Retract action slide handle as in (1) above; then release pressure on rear of action slide lock and push slide smartly forward to lock the breech bolt. Then attempt to retract the action slide handle. The action slide handle should not retract.

(3) Pull the trigger, thus allowing the hammer to move forward to the fired position and attempt to retract the action slide handle. The action slide handle should retract.

(4) Retract the action slide handle fully and then push forward until the breech bolt is fully forward but not raised to the locked position. Then pull the trigger to release the hammer. The trigger should not release the hammer until the breech bolt is fully locked.

(5) Place two or more dummy or fired shells in the magazine and work through the action to test gun for feeding, loading, extraction and ejection of shells. The second shell should not leave the magazine until the breech bolt is locking behind the first shell in the chamber.

NOTE: Fired shells will not work through the action as easily as live or dummy shells as they are somewhat deformed through being fired. Therefore, allowance should be made for friction and smoothness of action in positioning the shell.

(6) With breech bolt locked and hammer in cocked position, push the trigger (safety) lock all the way to the right and attempt to pull the trigger to release the hammer. The trigger should not pull nor the hammer release.

(7) Push trigger (safety) lock all the way to left so that red band shows and attempt to pull the trigger. The trigger should pull and hammer be released to fire the gun.

c. When gun does not operate and function smoothly and properly when tested as above, damaged or improperly assembled parts are indicated as follows:

(1) ACTION SLIDE STICKS. May be due to bent action slide bar, burs

WINCHESTER SHOTGUN, 12-GAGE, M12

on bar cam lug, or foreign matter in breech bolt camming aperture, burs on breech bolt guideways or guide.

(2) BREECH BOLT DOES NOT LOCK. May be due to foreign matter on face of bolt or on locking shoulder in top of receiver, or worn or burred action slide cam lug.

(3) HAMMER DOES NOT COCK PROPERLY, OR SLIPS. May be due to burs or foreign matter in sear notch in hammer, or burred sear nose, or weak or broken trigger spring.

(4) FIRING PIN DOES NOT RETRACT IN BREECH BOLT. May be due to burred or broken retractor or broken retractor spring.

(5) ACTION SLIDE LOCK DOES NOT FUNCTION. May be due to broken, or bent action slide lock springs, or improper assembly of hammer and lock in trigger guard, or burs on action slide bar. Refer to CAUTION following paragraph 3 q.

NOTE: Small lug on lower left side of hammer should lie between the two straight springs on the action slide lock when assembled.

(6) SHELL IS NOT EXTRACTED OR EJECTED. May be due to worn, burred or broken extractors, broken extractor springs, or broken ejector or ejector spring.

(7) TWO SHELLS FED INTO RECEIVER AT ONCE. May be due to broken, burred, or improperly assembled shell cut-off or foreign matter in cut-off seat in receiver.

(8) SHELL STICKS IN MAGAZINE. May be due to corroded or bent follower, dented tube, broken or kinked spring, or foreign matter in tube.

(9) TRIGGER (SAFETY) LOCK STICKS. May be due to burred lock spring plunger, broken plunger spring, or foreign matter in slots in lock.

d. Inspect barrel and test trigger pull (par. 3 n).

e. In addition to inspection for operation and functioning, the gun should be inspected generally for condition and defects noted. Attention should be directed to such defects as cracked wooden parts, cracked or deformed metal parts, dented magazine tube, loose screws or pins, loose or binding parts or assemblies, loose barrel or magazine, loose stock or butt plate, rust, dents, burs, or excessive wear of parts. If defects are such that early malfunction of the gun is indicated, the gun should be turned over to ordnance personnel for inspection and correction.

f. Where defects and malfunctions cannot be remedied by cleaning, lubrication, and simple adjustments of assembly, which lie within the scope of using troops, the gun should be turned over to ordnance personnel for a thorough inspection, correction and/or repair.

g. Removal of burs on working parts, trigger adjustments and like corrections should not be attempted by using troops as stoning of parts

SHOTGUNS, ALL TYPES

must be exacting, the angle of the faces concerned must not be changed, and volume of metal must not be materially reduced.

h. A loose barrel, which shakes when assembled, may be due to improper assembly caused by not inserting barrel far enough into receiver before mating interrupted threads of barrel and receiver, or it may be due to worn parts necessitating adjustment or replacement. If it is due to worn parts, the gun should be turned over to ordnance personnel for correction or repair.

i. If shell appears unnecessarily loose in chamber with breech bolt locked, the gun should be turned over to ordnance personnel to be checked for headspace.

j. Adjustment and maintenance of the gun, in the case of using troops, is limited to the removal and replacement of the parts and groups of parts as outlined in paragraphs 16 and 17, together with cleaning and lubrication and such adjustments as are necessary in assembling the gun as outlined.

19. CLEANING AND LUBRICATION.

a. Cleaning, oiling, and lubrication of this gun may be accomplished in a manner similar to that prescribed for the Winchester Gun M97 (par. 11). Attention should be given to corresponding parts and surfaces when lubricating. Barrel and magazine group should be removed from receiver of take-down gun when cleaning bore. Bore of solid-frame gun should be cleaned from the muzzle end, and a rag stuffed into the receiver to protect the action while cleaning (par. 3 i).

b. **Points to Lubricate Are:**
(1) Action slide bar opening in forward end of receiver.
(2) Action slide bar cam lug.
(3) Action slide lock pin.
(4) Carrier pivot.
(5) Breech bolt guide.
(6) Outer surface of magazine tube where action slide bears.
(7) Hammer pin.
(8) Shell cut-off pivot stud.

c. In very cold climates, oiling, and lubrication should be reduced to a minimum. Only surfaces showing signs of wear should be lightly oiled. Refer to "Special Maintenance," section XI.

Section IV

STEVENS SHOTGUN, 12-GAGE, M620A, M520, AND M620

	Paragraph
Description	20
Data	21
Operation	22
Functioning	23
Removal of groups	24
Replacement of groups	25
Field inspection	26
Cleaning and lubrication	27

20. DESCRIPTION.

a. Identification marks on these guns are generally to be found as follows:

(1) Name of maker and chamber length are stamped on the left side of the barrel near the breech end, and serial number of the gun in the operating handle bar slot in the left side of the barrel head.

(2) Name of maker and model number are stamped on the left side of the receiver and the serial number of the gun on the trigger plate tang (M520 and M620).

(3) On the M620A Gun, the serial number is stamped on the forward face of the trigger plate.

(4) On the M520 Gun (riot type), the letters U. S. are stamped on the left side of the receiver, and the serial number of the gun on the trigger plate tang.

b. The above guns (figs. 32, 33, 34, and 35) are of the same make but of different models, and are basically the same with regard to general description, operation, and functioning. Variations occur in details of design, principally in the trigger plate group. As the M620A Gun is of more recent design than the other two models it will be covered herein, with variations occurring in the other two models described as they occur.

c. This gun is a manually operated, repeating shotgun of the slide action, hammerless, solid-frame and take-down type. The gun is so constructed that the barrel, magazine, and operating handle group can easily be removed as a unit from the receiver by disengaging the magazine nut by screw action of the magazine tube to which it is threaded, and sliding the group downward out of the receiver. This construction facilitates cleaning and transportation (par. 3 i).

SHOTGUNS, ALL TYPES

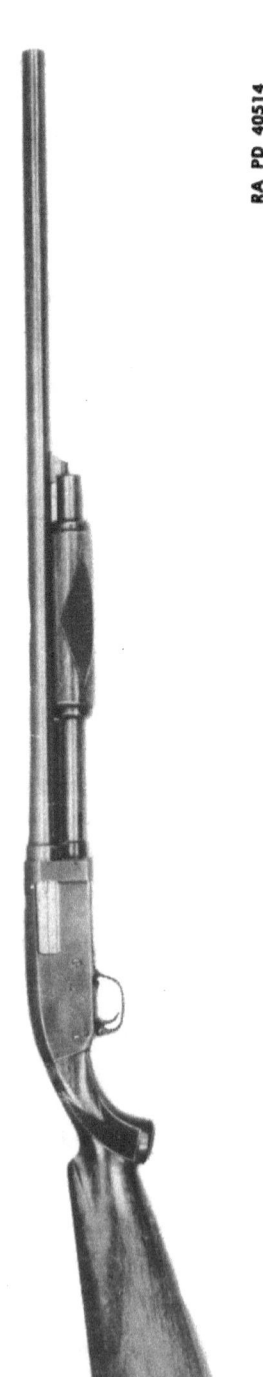

RA PD 40514

Figure 32 — Right Side View — Sporting Trap Type — Stevens Shotgun M620A

RA PD 40511

Figure 33 — Right Side View — Riot Type — Stevens Shotgun M520

TM 9-285
20

STEVENS SHOTGUN, 12-GAGE, M620A, M520, AND M620

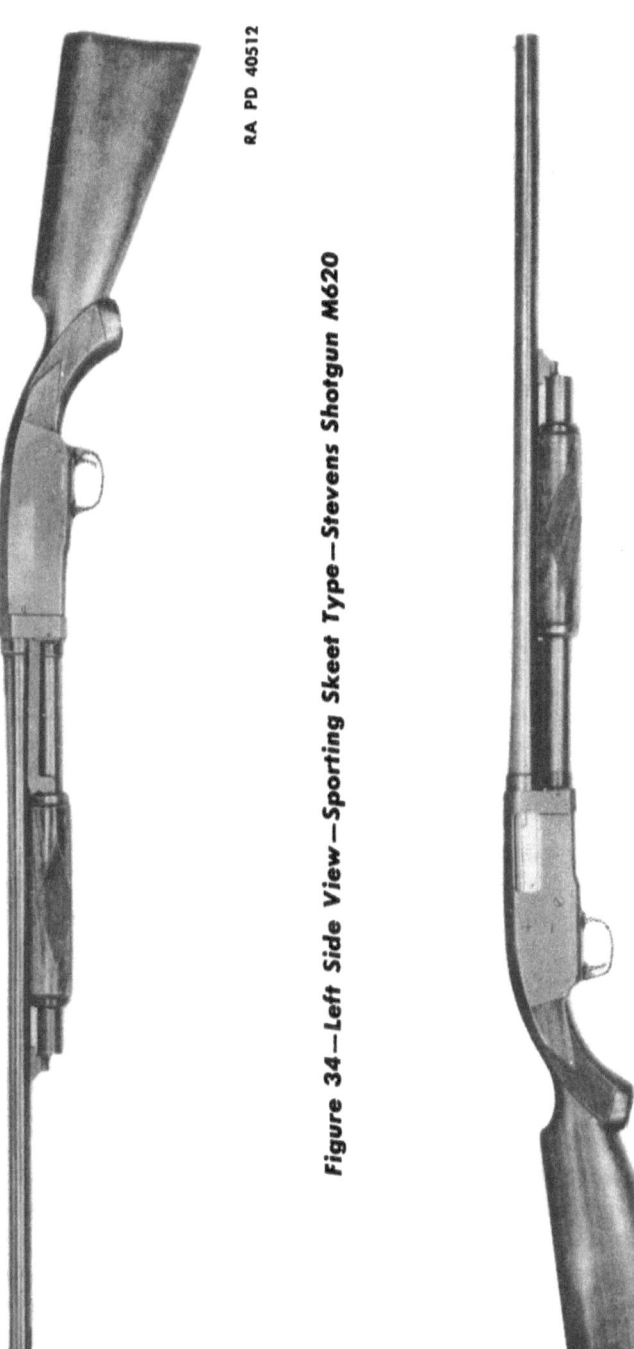

Figure 34—Left Side View—Sporting Skeet Type—Stevens Shotgun M620

Figure 35—Right Side View—Sporting Skeet Type—Stevens Shotgun M620

SHOTGUNS, ALL TYPES

d. This gun is furnished with barrels of different lengths and degrees of boring and may have other modifications of design. For convenience the guns (all models) will be classified as three types: riot, sporting skeet, and sporting trap, although variations of these types may occur.

(1) The riot-type gun (fig. 33) is usually furnished with a 20-inch plain barrel, bored full cylinder, and may be equipped with a bayonet attachment and hand guard attached to the muzzle end of the barrel and a leather sling attached to a sling swivel on the bayonet attachment and the stock. If equipped with bayonet attachment, it is solid-frame. (par. 3 i).

(2) The sporting skeet-type gun (figs. 34 and 35) is usually furnished with a 26-inch plain barrel, bored improved cylinder, and is without bayonet attachment or sling.

(3) The sporting trap-type gun (fig. 32) is similar to the skeet-type gun but is furnished with a 30-inch barrel, bored full choke and is without bayonet attachment or sling.

e. **General Description** (figs. 36, 37, 45, and 46).

(1) The stock of the gun is bolted to the tang which, in the M620A Gun, floats and locks to the receiver and trigger plate by the forward end of the tang seating in grooves in rear end of receiver and trigger plate, when assembled. In the M520 and M620 Guns, the stock is fastened to the receiver and trigger plate by a screw which passes vertically through an integral tang on the receiver, the stock grip, and screws into an integral tang on the trigger plate.

(2) The barrel, magazine, and operating handle bar group is attached to the forward end of the receiver by means of grooves and guides in the rear end of the barrel sliding into and mating with similar grooves and guides in the forward end of the receiver. When the group is in position, the magazine nut is moved rearward by screw action of the magazine tube, so that lugs on the nut engage in slots in the receiver and barrel head to hold the groups locked together. The magazine is of the tubular type with a capacity of five shells loaded end to end. The shells are pressed together and fed into the receiver by the force of the magazine spring acting upon the follower.

(3) The magazine tube is locked to the barrel at the rear by the magazine nut, and forward by a screw passing through a lug in the magazine plug and into a lug on the barrel. The operating handle is mounted and operated on the magazine tube, and the operating handle bar attached to the rear end of the operating handle tube passes through the forward end of the receiver and engages with and operates the slide. This slide in turn connects with and operates the sliding breech and lifter and cocks the hammer.

TM 9-285
STEVENS SHOTGUN, 12-GAGE, M620A, M520, AND M620

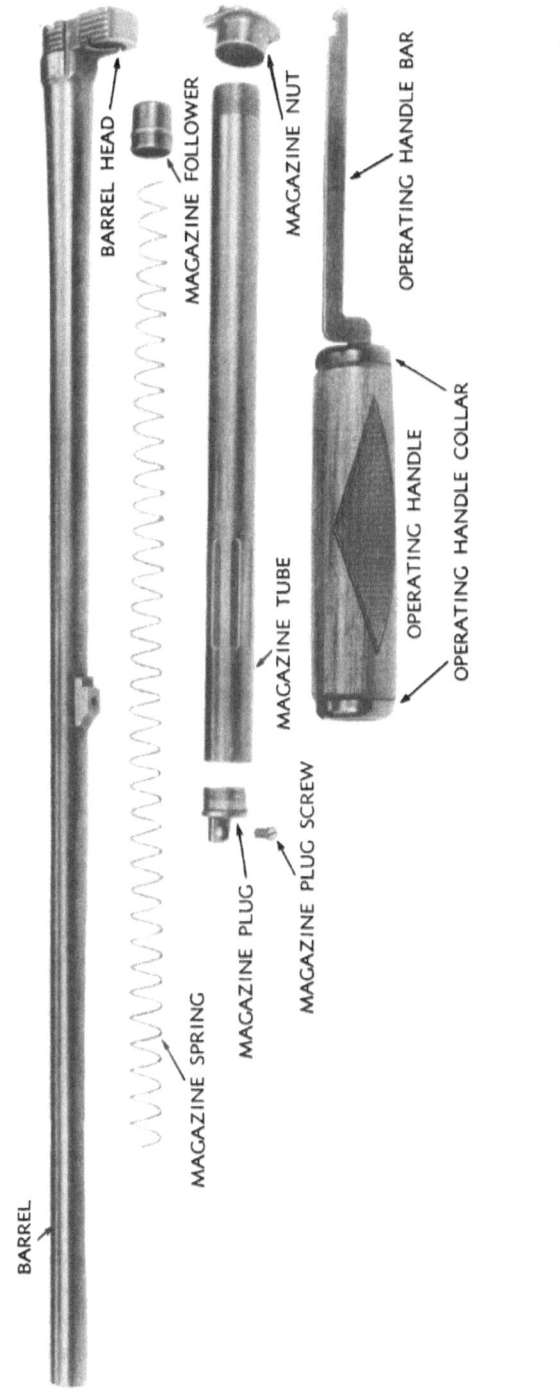

Figure 36 — Barrel, Magazine and Operating Handle Group — Disassembled View — Stevens Shotgun M620A

TM 9-285
20

SHOTGUNS, ALL TYPES

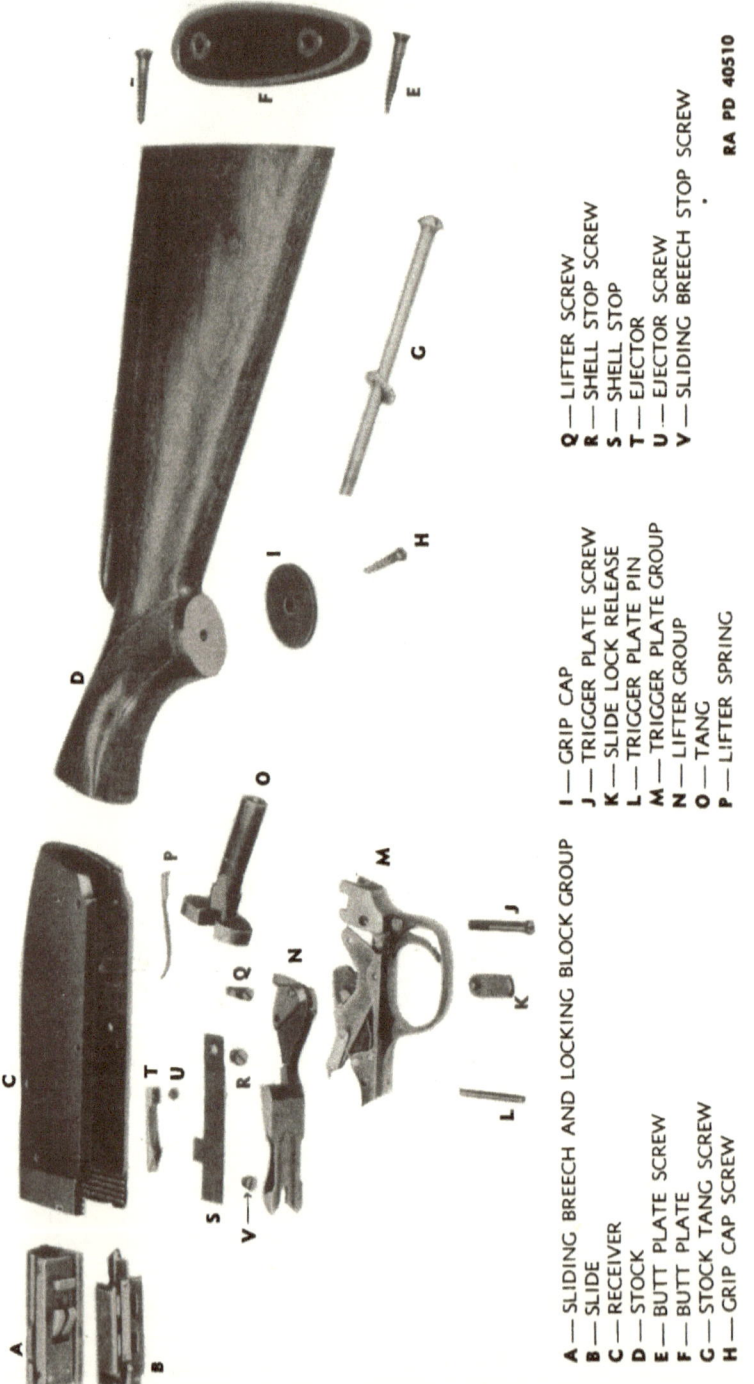

A — SLIDING BREECH AND LOCKING BLOCK GROUP
B — SLIDE
C — RECEIVER
D — STOCK
E — BUTT PLATE SCREW
F — BUTT PLATE
G — STOCK TANG SCREW
H — GRIP CAP SCREW
I — GRIP CAP
J — TRIGGER PLATE SCREW
K — SLIDE LOCK RELEASE
L — TRIGGER PLATE PIN
M — TRIGGER PLATE GROUP
N — LIFTER GROUP
O — TANG
P — LIFTER SPRING
Q — LIFTER SCREW
R — SHELL STOP SCREW
S — SHELL STOP
T — EJECTOR
U — EJECTOR SCREW
V — SLIDING BREECH STOP SCREW

RA PD 40510

Figure 37 — Stock and Receiver Group — Disassembled View — Stevens Shotgun M620A

STEVENS SHOTGUN, 12-GAGE, M620A, M520, AND M620

(4) The receiver contains the operating mechanism and to its lower rear end is attached the trigger plate to which is mounted the firing mechanism and the slide lock. The receiver is open at the bottom to permit loading, at the right side for ejection of the fired shell cases, and at the top for engagement with the locking block operating in the sliding breech. The lifter is pivoted to the inside of the receiver and operated by means of a spring functioned pawl on the rear end of the lifter engaging with a notch in the slide, together with the lifter spring pinned to the receiver at the rear of the lifter. The function of the lifter is to raise the shell to chamber level when released from the magazine.

NOTE: The receiver of the M520 and M620 Guns has an integral tang extending to the rear, while in the M620A Gun, the tang is floating as already described.

(5) The shell stop is screwed to the inside of the right wall of the receiver and is operated by the slide and spring action of the stop. The ejector is positioned in the left wall of the receiver and acts to kick the shell out of the ejection opening in the receiver when pulled to the rear out of the barrel chamber by the extractors positioned in the forward end of the sliding breech.

NOTE: The ejector in M620A and M520 Guns is fastened to the receiver by a screw, while that of the M620 Gun seats in an aperture in the receiver wall and is held in position by the sliding breech when assembled.

(6) The sliding breech (fig. 38) contains the extractors, right and left, the firing pin, and the locking block and their components. The locking block operates in radial grooves cut in the sliding breech and is operated by means of cam lugs on the lower end of the block, functioning with similarly shaped camming apertures in the slide, upon which the sliding breech rests. The slide moves the sliding breech back and forth and cams the locking block up and down in an aperture in the top of the receiver to lock and unlock the sliding breech. The extractors are of the usual claw type functioned by springs. The firing pin passes through the sliding breech and the locking block and is cammed back into the sliding breech on the rearward movement of the slide, by the locking block.

(7) The slide (fig. 38) moves in guideways in the receiver walls and supports, operates, locks, and unlocks the sliding breech, operates the lifter and shell stop, and cams back and cocks the hammer. The slide is in turn operated by the operating handle bar which engages with it by means of a lug on the bar engaging in a mating notch in the slide when assembled. The slide is blocked in the forward (locked) position by the slide lock mounted in the trigger plate. The slide and lock are held in

TM 9-285

SHOTGUNS, ALL TYPES

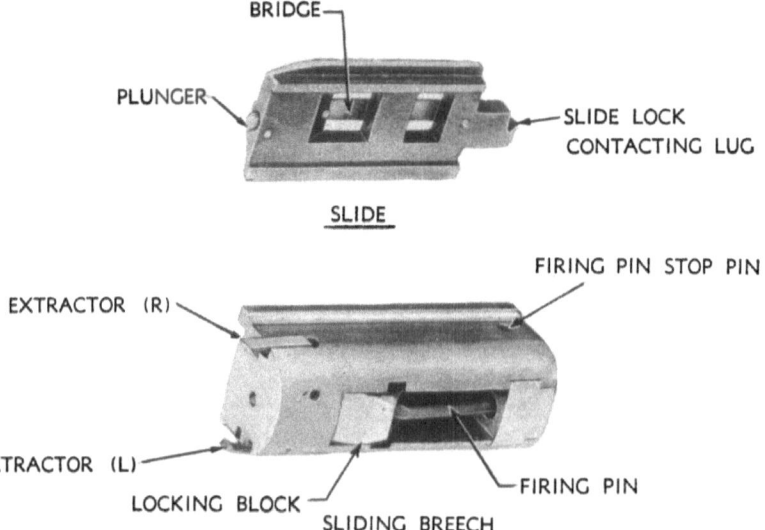

Figure 38—Slide and Sliding Breech—Front, Top and Right Side View—Showing Location of Parts—Stevens Shotgun M620A

close contact, in the locked position, by a spring plunger in the forward end of the slide, bearing upon the barrel head when assembled.

(8) The trigger plate group (fig. 39) contains the hammer, trigger, sear, slide lock, and safety, together with their springs and component parts.

NOTE: The trigger plates of the M520 and M620 Guns have an integral tang extending to the rear, while in the M620A Gun, the tang is floating as already described, and mates with both receiver and trigger plate when assembled. The principal differences in the trigger plate groups of the M620A (fig. 39), M520 and M620 Guns are as follows:

(a) Hammer. The hammers of the M620A and M620 Guns are the same except that the hammer of the M620 Gun has a small roller pinned to its lower end. The sear hook points upward in both models. The hammer of the M520 Gun is similar to the M620A Gun, without roller, except that the sear hook points down.

(b) Mainspring. The mainspring of the M620A and M520 Guns are of the "mousetrap" type, pivoted on a cross pin. The mainspring of the M620 Gun is a long, leaf type of spring fastened to the trigger plate tang by a screw. This spring is slotted to allow the hammer and sear to operate through the slot, and bears on the roller in the hammer.

TM 9-285

STEVENS SHOTGUN, 12-GAGE, M620A, M520, AND M620

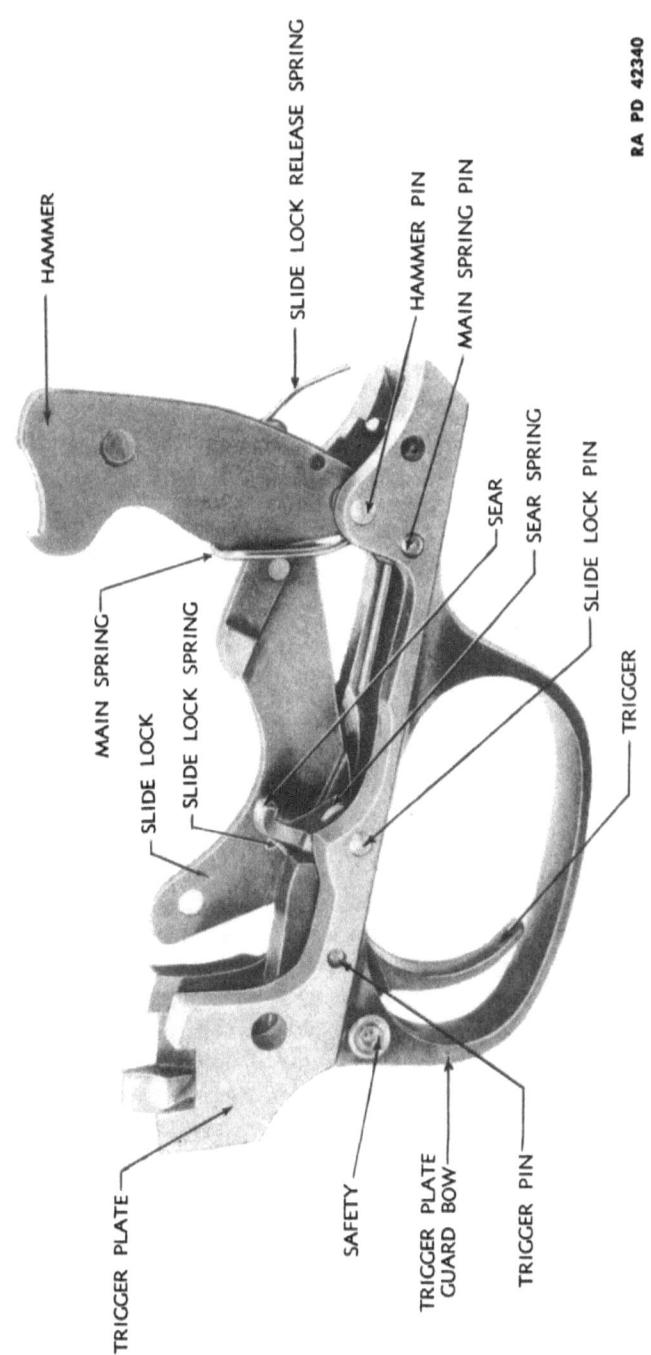

Figure 39 — Trigger Plate Group — Right Side View — Showing Location of Parts — Stevens Shotgun M620A

SHOTGUNS, ALL TYPES

(c) *Sear.* The sears of the M620A and M620 Guns are the same and pivot on the slide lock pin, and are functioned by a torsion spring. The sear of the M520 Gun is somewhat longer and seats in a slot in the top of the trigger and is functioned by a small coil spring seated between sear and trigger. Sear and trigger are pivoted on the trigger pin.

(d) *Trigger.* The triggers of the M620A and M620 Guns are the same, are pivoted on the trigger pin, and operate the sear by levering down the rear end. The trigger of the M520 Gun has a slotted top in which the sear and sear spring seat. When pulled, the trigger raises the rear end of the sear with it to release the hammer. In the M620A and M620 Guns the sear spring functions the trigger, while in the M520 Gun, the torsion trigger spring mounted on the trigger pin functions the trigger.

(e) *Slide lock.* The slide lock (figs. 43 and 44) of the M620A, M520, and M620 Guns are practically the same. In the M620A and M620 Guns, the slide lock spring is of the torsion type and mounted on the trigger pin and bears down upon the rear end of the slide lock. In the M520 Gun, this spring is of the safetypin type and seats in a hole in the top of the left wall of the trigger plate, and bears upward on the rear pin of the slide lock link positioned in the forward end of the slide lock.

(f) *Safety.* The safeties of the M620A and M620 Guns are of the cylindrical type, mounted in the rear of the trigger plate guard bow. The safety in the M520 Gun is of the lever type, the lever pivoted in the rear end of the trigger plate, and the thumbpiece in the tang of the receiver. The thumbpiece operates the lever when the trigger plate group is assembled to the receiver.

(9) The slide lock, pivoted in the trigger plate, engages with the slide to block its rearward movement after it has locked the sliding breech in position, thereby preventing premature unlocking of the sliding breech. The lock is disengaged from the slide either by the action of the slide lock release spring, functioned by the descending hammer with which it is engaged, or by manual pressure upward on the rear end of the slide lock through the medium of the slide lock release, which extends downward through the floor of the trigger plate to the left of the trigger.

(10) The safety (fig. 39) of the M620A and M620 Guns operates laterally in the rear end of the trigger plate guard bow to block and free the trigger, thus preventing or allowing its pulling to fire the gun. In the M520 Gun, the safety thumbpiece which acts upon the safety lever, positioned as described above, operates longitudinally to accomplish the same results.

STEVENS SHOTGUN, 12-GAGE, M620A, M520, AND M620

21. DATA.

Gage of bore	12
Boring of barrel—riot type	Cylinder
Boring of barrel—sporting skeet type	Improved cylinder
Boring of barrel—sporting trap type	Full choke
Type of action	Slide
Type of firing mechanism	Hammerless
Type of magazine	Tubular
Capacity of magazine	5 shells
Length of barrel—riot type	20 in.
Length of barrel—sporting skeet type	26 in.
Length of barrel—sporting trap type	30 in.
Length of stock and receiver (approx.)	20 in.
Length of assembled gun—riot type (approx.)	40 in.
Length of assembled gun—sporting skeet type (approx.)	46 in.
Length of assembled gun—sporting trap type (approx.)	50 in.
Weight of assembled gun—riot type (M520) (approx.)	7 lb
Weight of assembled gun—sporting skeet type (M620) (approx.)	7 5/8 lb
Weight of assembled gun—sporting trap type (M620A) (approx.)	7 5/8 lb

22. OPERATION.

a. The gun is operated by moving the operating handle fully and smartly backward and forward. This action unlocks the sliding breech, extracts and ejects the fired shell casing, cocks the hammer, transfers a live shell from the magazine to the chamber of the barrel, and relocks the sliding breech behind the shell.

CAUTION: During operation, the muzzle of the gun should always be pointed at a safe spot.

b. Before the operating handle can be retracted, the slide lock must be disengaged from the slide (fig. 40). To facilitate this, the operating handle should be moved slightly forward to allow the lock to disengage. The slide is held snugly in engagement with the slide lock through the force exerted by the slide spring plunger against the barrel head. If the gun has been fired and the hammer consequently forward, the only movement necessary is to reciprocate the operating handle as above as the descending hammer through the medium of the slide lock release spring, with which it is engaged, has already disengaged the slide lock from the slide. If, however, the hammer is in the cocked position, it is necessary to push up on the slide lock release which is visible on the left side of the trigger plate, just to the left of the trigger, before retracting the operating handle. When the gun is being fired as a repeater, the preliminary for-

SHOTGUNS, ALL TYPES

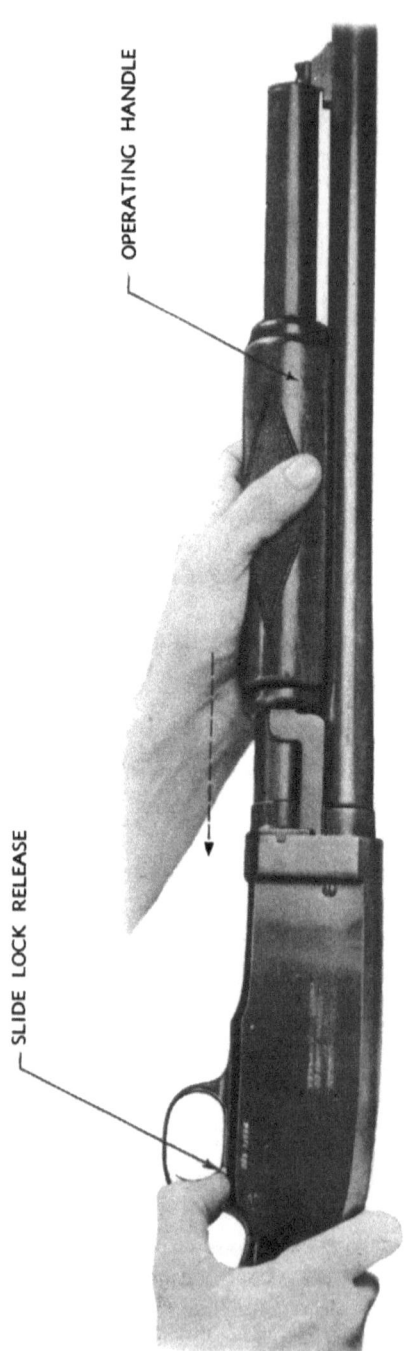

Figure 40—Disengaging Slide Lock and Retracting Operating Handle—Stevens Shotgun M620

STEVENS SHOTGUN, 12-GAGE, M620A, M520, AND M620

ward movement of the operating handle is accomplished by the recoil of the gun away from the operating handle, which is held by the operator.

CAUTION: During these operations, the finger should remain outside the trigger plate guard bow. Reciprocation should be full and smart to insure extraction of the shell, cocking of the hammer, and complete locking of the sliding breech and engagement of the slide lock. Slamming of the mechanism, however, should be avoided.

c. With the gun loaded, locked, and cocked, the only movement necessary to fire the gun is the pulling of the trigger. Pressure on the trigger should be removed while the gun is being operated.

d. The trigger safety (M620A and M620 Guns) (fig. 41), positioned in the rear of the trigger plate guard bow, operates laterally. When pushed to the left, the trigger is free to be pulled and the gun fired. When pushed to the right, the trigger is blocked and cannot be pulled nor the gun fired. In the M520 Gun, the safety thumbpiece is positioned in the tang of the receiver and operates longitudinally. When pushed forward, the trigger is free to be pulled and the gun fired. When pulled to the rear, the trigger is blocked and cannot be pulled nor the gun fired. With a shell in the chamber, it is best to block the trigger by pushing the safety to the right (M620A and M620 Guns) or to the rear (M520 Gun) unless the gun is to be fired immediately.

e. **To Load and Unload the Magazine.**

(1) To load the magazine (fig. 42) press a shell, nose first, against the lifter and into the rear end of the magazine against the magazine follower until it slips in front of and is retained by the lifter. Load another shell in the same way by pressing it against the lifter and base of the first shell loaded, until five in all are loaded. Loading should be done with the sliding breech closed and locked.

(2) To unload the magazine, press the carrier up into the receiver until free of the base of the shell in the magazine and allow the shells to slip out one by one. Sliding breech must be in locked position.

f. **To Load and Unload the Gun.**

(1) To load a shell from the magazine into the chamber, slide the safety to safe position (par. 22 d). Then disengage slide lock and reciprocate the operating handle (par. 22 b). Another shell may then be loaded into the magazine. Test locking of sliding breech by attempting to retract operating handle. Operating handle should not retract. Allow safety to remain in safe position unless gun is to be fired immediately.

(2) To unload the gun, slide safety to safe position and, unload the magazine as explained above. Then disengage slide lock and retract operating handle to extract and eject shell in chamber. Inspect magazine and chamber to make sure gun is fully unloaded.

TM 9-285
22

SHOTGUNS, ALL TYPES

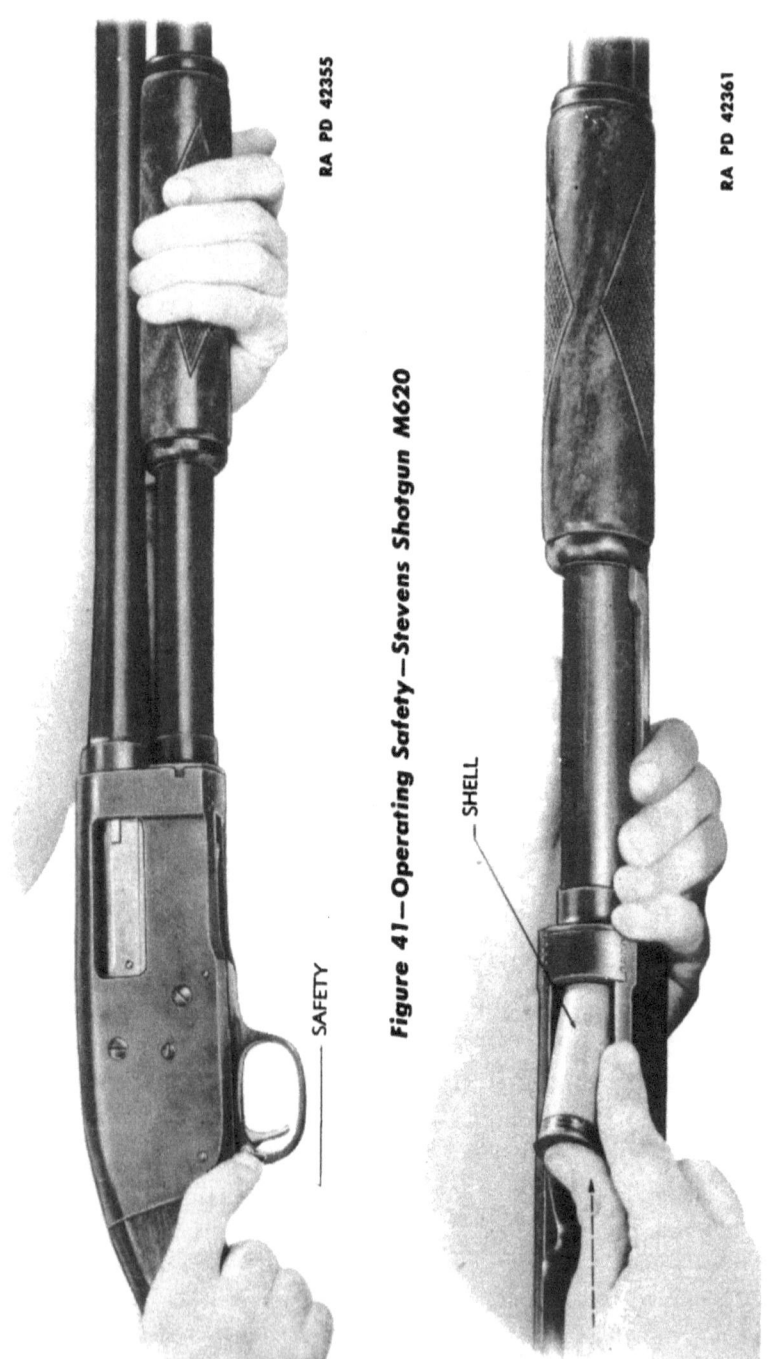

Figure 41 — Operating Safety — Stevens Shotgun M620

Figure 42 — Loading Magazine — Stevens Shotgun M620

STEVENS SHOTGUN, 12-GAGE, M620A, M520, AND M620

g. *To Load the Chamber Only.* With the magazine empty, slide safety to safe position as explained above, retract the sliding breech by pulling the operating handle fully to the rear as explained, and push a shell directly into the chamber. Then push the operating handle fully and smartly forward to lock the sliding breech behind the shell. Test locking of slide (par. 22 f).

23. FUNCTIONING.

a. As already explained, the functioning of the operating mechanism is accomplished by the reciprocation of the operating handle. The rear end of the operating handle bar is engaged with the slide by means of a lug on the bar mating with a notch in the slide. As the operating handle bar is moved to the rear, it pushes the slide to the rear. The locking block, which is a radially curved piece, passes through the sliding breech, which rests upon the slide, and the top of the receiver, when in the locked position, to lock the sliding breech to the receiver. When in this position, the lower end of the locking block rests upon the slide which thus holds it up in position. As the slide moves to the rear, it cams down the locking block from the top of the receiver by means of the lugs on the lower end of the locking block slipping into and engaging with corresponding camming apertures in the slide. The locking block in turn cams back the firing pin which passes through it, from the face of the sliding breech. As the slide moves to the rear, it pulls the sliding breech with it by means of the locking block. The sliding breech extracts the fired shell from the chamber by means of the extractors, and the shell is knocked out of the receiver, in passing, by the ejector, positioned in the left side of the receiver wall. As the slide moves forward again, it pulls the sliding breech with it, and at the end of the forward movement, cams up the locking block through the sliding breech and into the opening in the top of the receiver to lock the sliding breech to the receiver.

b. During the rearward movement, the slide cams back and rides over the hammer, mounted in the trigger plate. When the hammer reaches the extreme rearward position, it is caught and held by a hook on the slide lock engaging with a lug on the side of the hammer. The hammer is thus locked back, disengaged from the sear, until released from the slide lock by the slide, acting upon the slide lock during its forward movement. The hammer when released is caught and held by the sear, and the slide lock springs up behind the slide to lock it in position as explained below.

c. When the slide lock (figs. 43 and 44) is disengaged from the slide either by manual pressure or by the descending hammer (par. 20 e (9)), and the operating handle retracted, the slide moves to the rear, riding over the forward end of the lock. As the slide returns on the forward

SHOTGUNS, ALL TYPES

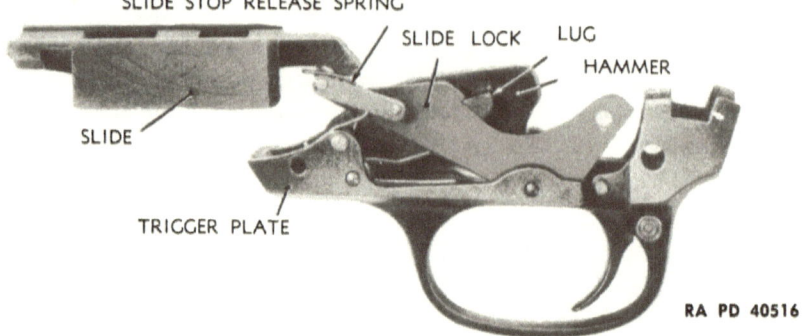

Figure 43 — Slide and Trigger Plate Groups — Slide Riding Over Lock — Hammer Locked — Stevens Shotgun M620A

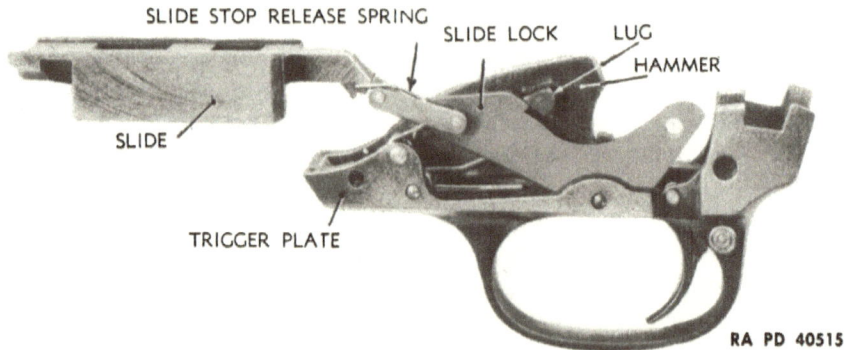

Figure 44 — Slide and Trigger Plate Groups — Slide Blocked — Hammer Unlocked — Stevens Shotgun M620A

movement, and nears the forward position, a lug on the rear of the slide cams the forward end of the lock down to release the hammer, which moves forward and is caught by the sear as above. Immediately after camming down the forward end of the lock, the slide moves from engagement with the lock and locks the sliding breech. The lock thus freed, springs up behind the slide to block its rearward movement and thus prevents premature unlocking of the sliding breech.

d. As the slide is pulled forward, it operates the lifter rapidly up and down by means of a spring functioned pawl on the lifter engaging in a notch in the slide. The lifter is returned to the lowered position by the lifter spring positioned in the receiver. The lifter lifts the shell resting upon it up into line with the chamber and the sliding breech, and then returns in time to catch the next shell released from the magazine by the action of the slide on the shell stop. The shell, thus raised to line of

STEVENS SHOTGUN, 12-GAGE, M620A, M520, AND M620

chamber, is pushed forward into the chamber by the sliding breech as it moves forward with the slide. The slide depresses and disengages the shell stop as it reaches the end of the forward movement, thereby releasing a shell from the magazine which is driven into the receiver upon the lifter by the force of the magazine spring.

e. The hammer is released to fire the gun by pulling the trigger which levers the sear from engagement with the hammer.

24. REMOVAL OF GROUPS (figs. 36, 37, 45 and 46).

a. As all three models of this gun are practically the same except for the trigger plate group and minor changes in design the M620A Gun (latest model) will be covered herein and the variations in procedure in the case of the other models noted as they occur. Groups and parts should be removed and replaced in the order given below. Groups and parts, when removed, should be placed on a clean, flat surface and care observed to prevent loss of screws and small parts. Remove as follows:

(1) BAYONET ATTACHMENT, BARREL, MAGAZINE, AND OPERATING HANDLE GROUP.

(a) The bayonet attachment may be removed from the barrel in a manner similar to that prescribed for the Winchester Gun M97 (par. 8 a (1) (a)).

(b) With bayonet attachment removed, hold gun bottom side up and press slide lock release which extends through the trigger plate to the left of the trigger. Then push operating handle forward slightly to allow lock to disengage, and then pull handle to the rear as far as it will go to retract the sliding breech. If sliding breech is in forward position, the extractors will interfere with removal of barrel.

(c) Turn magazine tube clockwise by means of milled surface until lugs on lock nut are out of mating notches in receiver (fig. 47).

(d) Grasp receiver in one hand and barrel in the other, with hands close together, and slide the barrel head upward out of the receiver about 1/8 inch, or as far as it will go. This disconnects the slide handle bar from the slide (fig. 48). Then draw operating bar handle forward as far as it will go until rear end of bar clears receiver (fig. 49).

(e) Slide barrel, magazine, and operating handle group upward out of receiver (fig. 50).

(2) SLIDE AND SLIDING BREECH GROUP.

(a) Remove sliding breech stop screw, located forward in left wall of receiver by unscrewing counterclockwise.

(b) If sliding breech is in the locked position, press up on slide lock release to disengage lock from slide, and move slide to rear until sliding breech is unlocked from receiver. Then move sliding breech together with slide out of forward end of receiver. (If slide is moved forward, it

TM 9-285
24

SHOTGUNS, ALL TYPES

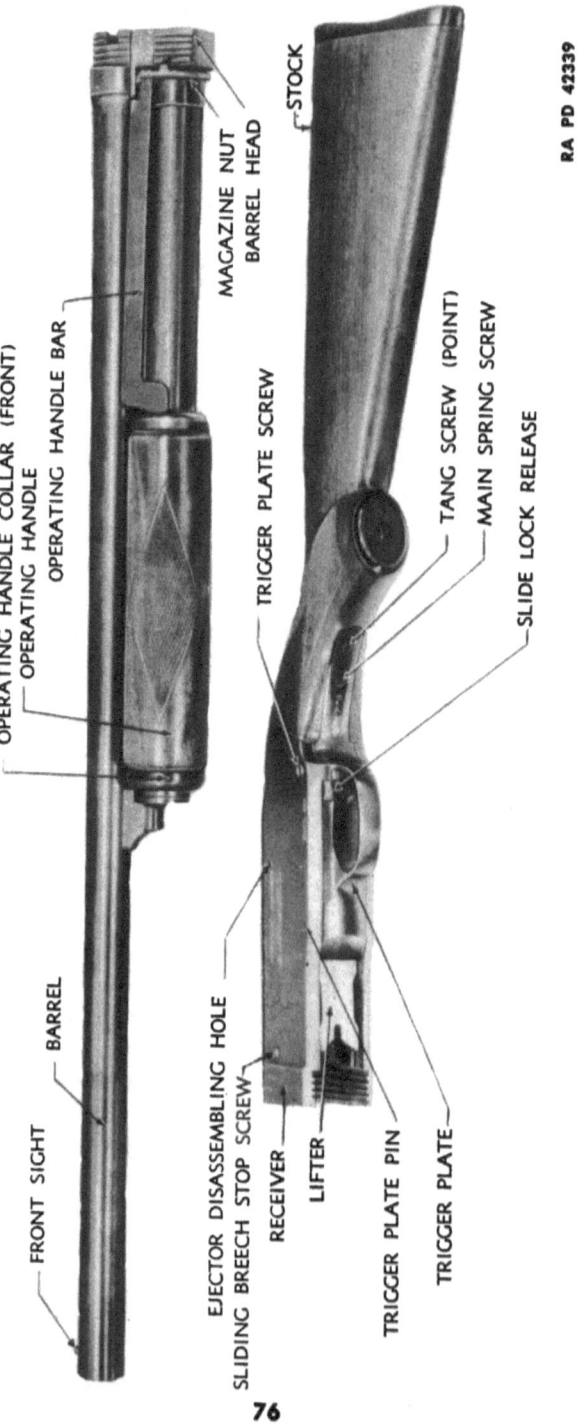

Figure 45 — Gun Taken Down — Left Side View — Showing Location of Parts — Stevens Shotgun M620

TM 9-285

STEVENS SHOTGUN, 12-GAGE, M620A, M520, AND M620

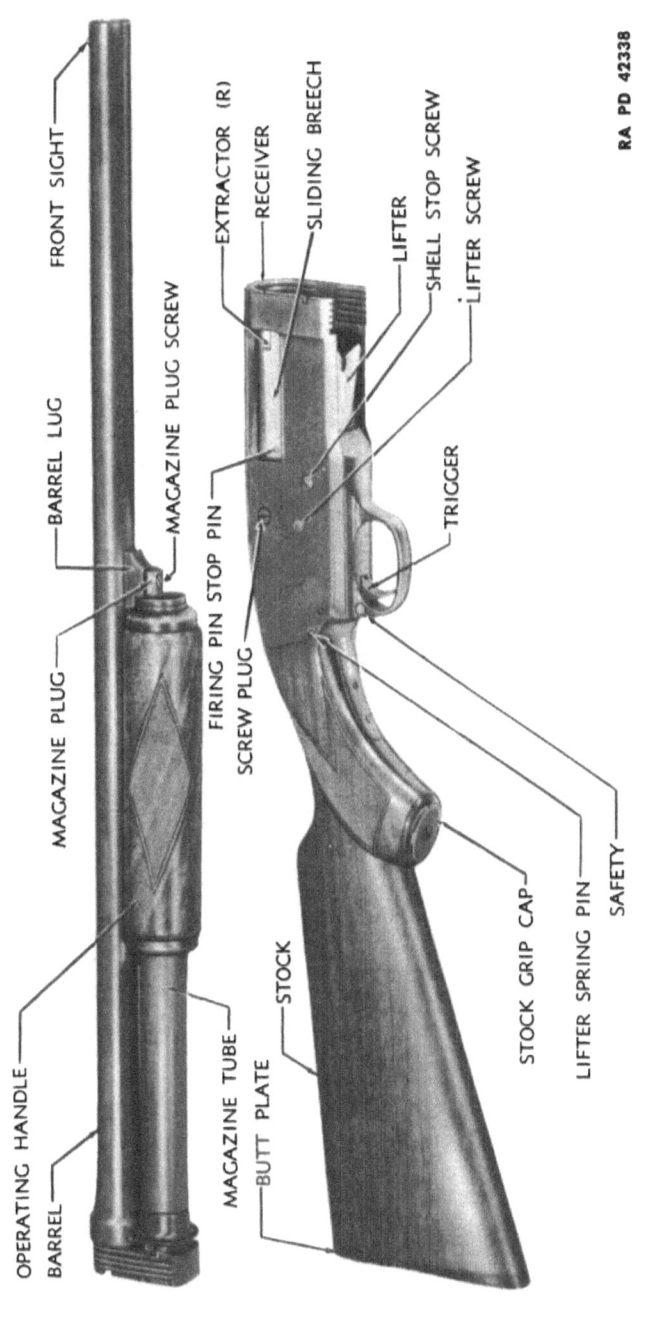

Figure 46 — Gun Taken Down — Right Side View — Showing Location of Parts — Stevens Shotgun M620

SHOTGUNS, ALL TYPES

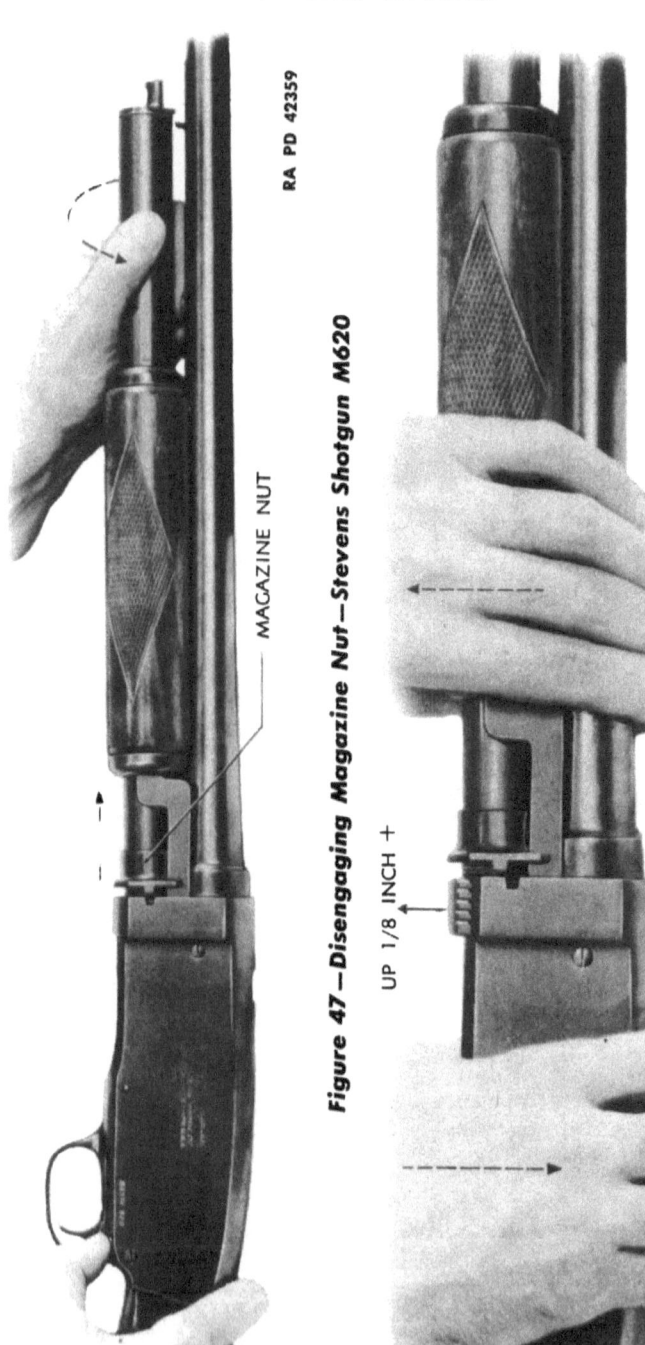

Figure 47 — Disengaging Magazine Nut — Stevens Shotgun M620

Figure 48 — Disengaging Operating Handle Bar from Slide — Stevens Shotgun M620

TM 9-285

STEVENS SHOTGUN, 12-GAGE, M620A, M520, AND M620

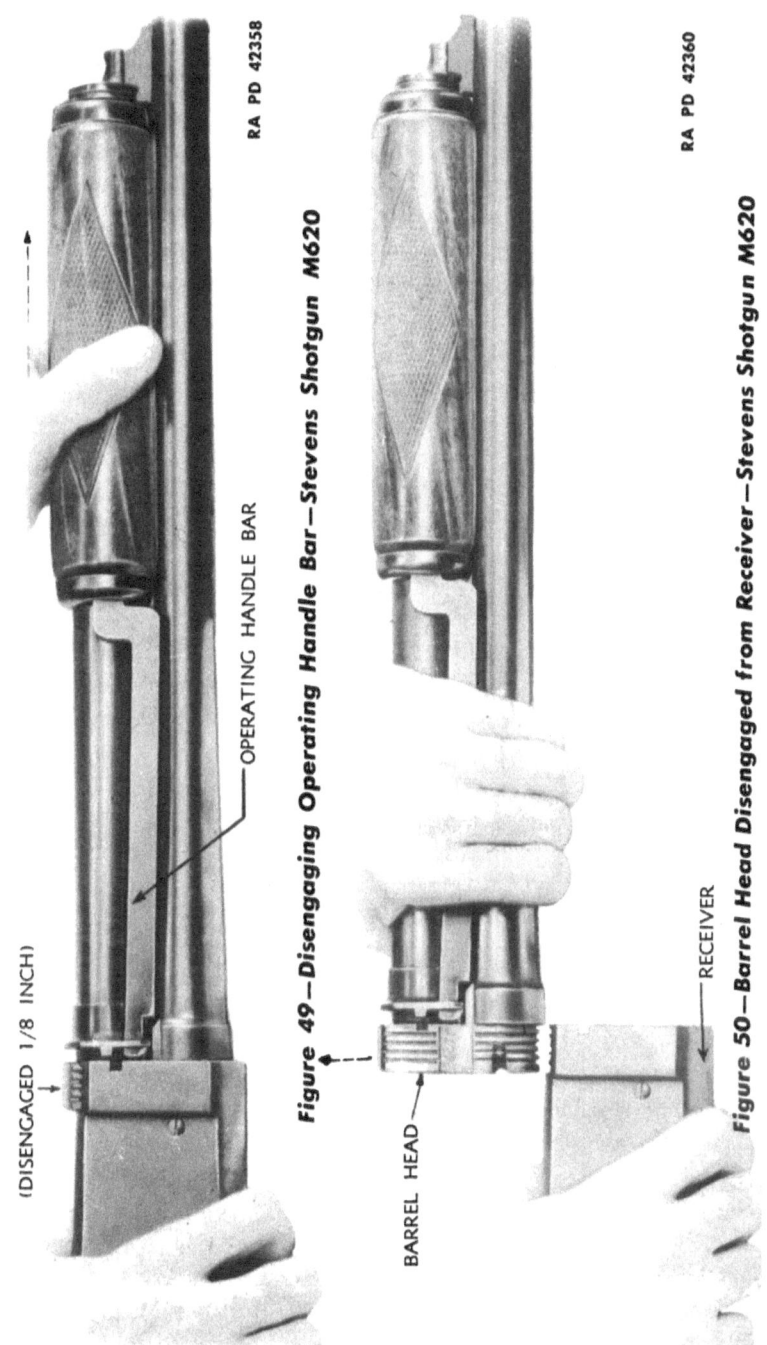

Figure 49 — Disengaging Operating Handle Bar — Stevens Shotgun M620

Figure 50 — Barrel Head Disengaged from Receiver — Stevens Shotgun M620

SHOTGUNS, ALL TYPES

will lock the sliding breech, and thus prevent removal.) If sliding breech is in the rearward position, it may be moved forward by lightly tapping end of receiver on a wooden block. The slide may now be lifted from the sliding breech.

NOTE: The ejector of the Model 620A and 520 Guns is fastened to the inner left wall of the receiver by a screw and should not be removed. The ejector on the Model 620 Gun seats in a countersink in the same position. If loose, it should be removed to prevent loss by inserting small drift in hole in left wall of receiver, from the outside, and pushing ejector from its seat, into the receiver. Observe care against loss of ejector spring.

(3) STOCK. In order to remove the trigger plate and lifter groups from the receiver the stock must first be removed as follows:

(a) Remove the two butt plate screws by unscrewing counterclockwise and remove butt plate.

(b) With long shanked screwdriver, unscrew stock tang screw in bottom of hole, bored in butt end of stock, by turning counterclockwise until disengaged from tang.

(c) Pull stock to rear from engagement with receiver and tang. Tang screw and washer may then be removed from rear end of stock.

NOTE: To remove stock from the Model 520 and 620 Guns, unscrew tang screw from top of receiver tang, by turning counterclockwise. Remove screw and pull stock to the rear from receiver. If removal is difficult, loosen trigger plate screw.

(4) TRIGGER PLATE GROUP. The hammer should be in the cocked position when removing trigger plate group, in order to minimize chance of damage to slide lock release spring.

(a) With stock removed, remove trigger plate screw from rear left side of receiver by turning counterclockwise. Then with straight pin drift, punch out trigger plate pin located about three inches forward of trigger plate screw.

(b) Pull trigger plate downward out of receiver. The floating tang (model 620A only) can then be pulled down out of receiver. Be careful not to lose the slide lock release when removing trigger plate.

CAUTION: Observe care when removing groups not to catch end of slide lock release spring, projecting from forward end of slide lock, on rear of the lifter. If this spring becomes deformed, the slide lock will not function. When the hammer is in rearward (cocked) position, the end of spring is less liable to catch. The function of the slide lock spring (M620A and M620 Guns) is to depress the rear end of the slide lock, thus raising the forward end, either to engage the hook on the lock with the lug on the hammer on the rearward movement or to engage the lock behind the slide to block it on the forward movement. The pressure

STEVENS SHOTGUN, 12-GAGE, M620A, M520, AND M620

exerted on the forward end of the slide lock by the slide lock release spring determines whether or not the slide lock spring has enough force to function properly. If the slide lock release spring becomes deformed and bears too heavily on the forward end of the lock, the slide lock spring will not be strong enough to counteract this excess pressure and raise the forward end of the lock when necessary. Thus the hammer will not be locked back and/or the slide blocked. If the spring does not bear heavily enough, it will not lever the nose of the lock down when the hammer moves forward. The torsion force of the slide lock release spring as well as the shape and position of its forward end, determine the pressure exerted at the correct time. The slide lock spring of the M520 Gun bears upward on the forward end of the slide lock.

(5) LIFTER.

(a) With the trigger plate group removed, slide lifter inward off its screw and remove from receiver.

(b) If lifter spring positioned in receiver to rear of lifter is loose, it should be removed to prevent loss by pushing in on pin end, shown in extreme right rear, outside face of receiver, using a small pin drift. Otherwise it should not be removed.

25. REPLACEMENT OF GROUPS.

a. Groups and parts should be thoroughly cleaned, lightly oiled, and gun lubricated, where necessary, before replacing groups. Replace as follows:

(1) LIFTER.

(a) If the lifter spring has been removed, replace by inserting pin on rear end into pin hole in extreme right rear inner wall of receiver. Press pin through hole using care to see that spring lies flush in its groove when seated. Pin should project through wall of receiver until flush. If pin is loose in hole, it can be expanded slightly by tapping outside center of pin lightly with a small prick punch.

(b) With receiver bottom side up and barrel opening to left, grasp lifter by forked end and insert vertically into receiver so that pawl on rear end points to forward end of receiver. Slide lifter onto its screw, positioned in the inner right wall of receiver and rotate forked end forward, at the same time pressing lifter in towards wall of receiver until lug on right side of lifter engages under lifter spring. Then, press lifter in against receiver wall and rotate forked end forward and down. If trigger plate is not to be replaced immediately, insert pin in trigger plate pin hole in receiver to retain lifter.

(2) TRIGGER PLATE GROUP.

(a) With receiver bottom side up, barrel opening to left, and lifter

SHOTGUNS, ALL TYPES

assembled to receiver, insert oval end of tang into the groove in rear top of receiver, with stem to rear, so that the relief cut in tang faces towards left side of the receiver.

(b) Cock hammer and position slide lock release in rear of slide lock so that stud on release fits in hole in lock and release lies flush with left face of trigger plate.

(c) Insert trigger plate, with trigger to rear, into bottom of receiver at an angle so that forward end of slide lock will slip under rear end of lifter. Then, press rear end of trigger plate down towards receiver until trigger plate screw hole in trigger plate is about ⅛ inch ahead of corresponding hole in receiver. Then, press trigger plate into receiver and to rear until tang groove in rear of trigger plate mates with bottom of tang oval while top of tang oval is still seated in its groove in receiver. When this mating is effected, press trigger plate down and to rear into receiver until it is flush with receiver, and screw and pin holes aline.

CAUTION: Care should be observed to prevent forward end of slide lock release spring from catching in carrier during replacement of trigger plate group and becoming deformed. If this spring becomes deformed, the slide lock will not function. If hammer is forward during replacement, end of spring is forced farther forward and is more liable to catch. Refer to CAUTION, paragraph 24 a (4) *(b)*.

(d) Insert trigger plate screw into its counterbored hole in rear left side of receiver and screw in part way. Then, insert trigger plate pin in its hole, just ahead of screwhole, rounded end first, and drive through receiver and trigger plate until flush with both sides of receiver. Then, tighten trigger plate screw.

NOTE: To replace trigger plate group of Model 520 and 620 Guns, slide trigger plate into receiver as explained above. Replace trigger plate screw and screw in part way. Then replace trigger plate pin. After stock has been replaced, draw trigger plate screw down tightly. This will allow the stock to slip between the two tangs more easily. Be sure rear end of safety lever of the M520 Gun is engaged with thumbpiece in receiver tang before seating trigger plate group.

(3) STOCK.

(a) With lifter and trigger plate assembled to receiver, slide stock on stock tang, pistol grip facing down. Mate tenon of stock with mortised rear end of receiver so that stock fits flush with rear of trigger plate and receiver.

(b) Place washer on stock tang screw and insert screw in hole in rear of stock. Mate threads in screw and tang; then screw down clockwise, drawing stock and receiver tightly together.

(c) Replace butt plate and butt plate screws. Do not draw butt plate

STEVENS SHOTGUN, 12-GAGE, M620A, M520, AND M620

screws down too tightly as threads in stock may strip out or composition butt plate crack.

NOTE: To replace stock on Model 520 and 620 Guns, insert grip end of stock between tangs on receiver and trigger plate and push towards receiver until tenon of stock grip mates with mortise in rear end of receiver, and tang screw holes in both tangs and stock grip aline. Insert tang screw through receiver tang and stock grip, thread into trigger plate tang and draw down tightly. Do not force assembly. If stock grip will not slide into place easily, loosen trigger plate screw, seat stock; insert tang screw, and then tighten trigger plate screw before tang screw is drawn down tightly. See that stock and receiver are rigid.

(4) SLIDE AND SLIDING BREECH GROUP.

(a) If the ejector (M620 Gun) has been removed from the receiver, it should be replaced before replacing the slide and sliding breech.

(b) Place the slide on the bottom of the sliding breech so that the cam lugs of the locking block seat in the lug camming apertures in the slide, and the lug on the rear of the slide is facing in the opposite direction to the extractors on the sliding breech.

(c) Move the slide to the rearward position with respect to the sliding breech so that the locking block is flush with the top of sliding breech. Then, insert slide and sliding breech, extractors facing out, into the forward (open) end of the receiver with sliding breech towards top of receiver and guides on slide mating with guideways in inner walls of receiver.

(d) Move the slide and sliding breech together towards the rear of the receiver until the slide strikes the shell stop. Press stop into its groove in the receiver and move the slide and sliding breech past it until rear of the slide strikes the slide lock. Press upward on slide lock release to disengage lock and move slide and sliding breech to rear. Then replace sliding breech stop screw. The slide should be moved and not the sliding breech, in order to keep locking block in the lowered position so as not to interfere with receiver.

(5) BAYONET ATTACHMENT, BARREL, MAGAZINE, AND OPERATING HANDLE GROUP.

(a) If sliding breech is in locked position, press upward on slide lock release, to disengage lock from slide. Then, move the slide to the rear to unlock the sliding breech and then move slide and sliding breech to rearward position. If sliding breech is in the forward position, the extractors will extend into the barrel aperture in the receiver, and the barrel cannot be inserted fully into the receiver.

(b) With the operating handle pushed fully forward and receiver and barrel group bottom side up, grasp barrel in one hand and receiver in the other, with the hands close together, and insert the mating guides of

SHOTGUNS, ALL TYPES

the barrel head into the mating grooves in the receiver, and slide barrel head down into the receiver until barrel head is flush with bottom of receiver.

(c) Grasp operating handle and pull it to the rear until bar snaps into position (mates with slide); then turn magazine tube counterclockwise by means of the milled surface until the lugs on the lock nut enter the notches on sides of the receiver and barrel head. Then screw down tightly until nut is flush with forward face of receiver.

(d) See that magazine tube is turned in tight, and then push operating handle forward until action is locked. Then operate gun to test assembly.

(e) The bayonet attachment, if removed, may be replaced in a manner similar to that prescribed for the Winchester Gun M97 (par. 9 a (6)).

26. FIELD INSPECTION.

a. With the gun completely assembled, test the mechanism for proper functioning. Fired shells may often be used for testing, when dummy shells are not available, by turning in the uncrimped end so that the length of the shell will approximate that of a live shell. Use of live shells for testing is prohibited.

CAUTION: Be sure gun is fully unloaded before inspection.

b. Operate the gun as follows:

(1) With the sliding breech locked and hammer cocked, push operating handle slightly forward and press upward on the slide lock release showing at left of trigger. Pull operating handle fully and smartly to the rear and then push fully and smartly forward. Reciprocate operating handle thus several times to test smoothness of action.

(2) Retract operating handle as in (1) above; then release slide lock release and push operating handle smartly forward to lock the sliding breech. Then attempt to retract the operating handle. The operating handle should not retract.

(3) Pull the trigger, thus allowing the hammer to move forward to the fired position, and attempt to retract the operating handle. The operating handle should retract.

(4) Retract the operating handle fully and then push forward until the sliding breech is fully forward but not locked and the locking block not fully engaged in the aperture in the top of the receiver. Then pull the trigger to release the hammer. The hammer should not be released until the sliding breech is fully locked and the locking block fully seated in the aperture in top of the receiver.

(5) Place two or more dummy or fired shells in the magazine and work through the action to test the gun for feeding, loading, extraction and ejection of shells. The second shell should not leave the magazine

STEVENS SHOTGUN, 12-GAGE, M620A, M520, AND M620

until the first shell has been loaded into the chamber, and the sliding breech is locking behind it.

NOTE: Fired shells will not work through the action as easily as live or dummy shells as they are somewhat deformed through being fired. Therefore allowance should be made for friction and smoothness of action in positioning the shell.

(6) With sliding breech locked and hammer in cocked position, slide trigger safety all the way to the right (M620A and M620 Guns) or to rear (M520 Gun) and attempt to pull the trigger. The trigger should not pull nor the hammer release.

(7) Slide trigger safety all the way to the left (M620A and M620 Guns) or forward (M520 Gun) and pull the trigger. The trigger should pull and the hammer be released to fire the gun.

c. When the gun does not operate and function smoothly and properly when tested as above, damaged or improperly assembled parts are indicated as follows:

(1) OPERATING HANDLE STICKS. May be due to bent operating handle bar, burs in bar slot in receiver or barrel head, or dented magazine tube.

(2) SLIDING BREECH DOES NOT LOCK. May be due to foreign matter on face of sliding breech, in extractor grooves in barrel head or in locking block aperture in sliding breech, broken firing pin, or burs on edges of locking block or locking aperture in top of receiver.

(3) HAMMER DOES NOT COCK PROPERLY OR SLIPS. May be due to burs or foreign matter on sear hooks of hammer or sear, worn or broken hooks, missing improperly assembled or broken sear or trigger springs, or improperly assembled or broken mainspring.

(4) FIRING PIN DOES NOT RETRACT INTO SLIDING BREECH. May be due to broken firing pin, or burs on firing pin or locking block.

(5) SLIDE DOES NOT GO FULLY FORWARD. May be due to jammed locking block or foreign matter in apertures in slide, jammed slide plunger, or broken firing pin.

(6) SLIDE LOCK DOES NOT FUNCTION. May be due to bent or burred slide lock, improperly assembled or bent slide lock spring, improperly assembled or bent slide lock release spring (usually the case), or burs on rear of slide. Refer to CAUTION, paragraph 24 a (4) *(b)*.

(7) SLIDE DOES NOT RETRACT FULLY. May be due to broken hammer, improperly assembled mainspring (M620 Gun), improperly assembled stock tang (M620A Gun), or broken firing pin.

(8) SHELLS ARE NOT EXTRACTED OR EJECTED. May be due to worn, broken or burred extractors, broken or missing extractor springs, or worn, broken, or missing ejector or ejector spring, or improperly assembled ejector.

SHOTGUNS, ALL TYPES

(9) TWO SHELLS FED INTO RECEIVER AT ONCE. May be due to bent, sticking, or broken shell stop or loose shell stop screw.

(10) SHELL STICKS IN MAGAZINE. May be due to corroded or bent magazine follower, dented tube, broken or kinked spring, or foreign matter in magazine tube.

(11) TRIGGER SAFETY STICKS. May be due to burred safety or trigger web (M620A and M620 Guns), broken thumbpiece spring or spring plunger (M520 Gun), or foreign matter in mechanism.

d. Inspect barrel and test trigger pull (par. 3 n).

e. In addition to inspection of the gun for operation and functioning, the gun should be inspected generally for condition and defects noted. Attention should be directed to such defects as cracked wooden parts, cracked or deformed metal parts, dented magazine tube, loose screws, loose or binding parts or assemblies, loose barrel or magazine, loose stock or butt plate, rust, dents, burs, or excessive wear of parts. If defects are such that early malfunction of the gun is indicated, the gun should be turned over to ordnance personnel for inspection and correction.

f. Where defects and malfunction cannot be remedied by cleaning, lubrication, and simple adjustments of assembly which lie within the scope of using troops, the gun should be turned over to ordnance personnel for a thorough inspection, correction, and/or repair.

g. Removal of burs on working parts, trigger adjustments, and like corrections should not be attempted by using troops as stoning of parts must be exacting, the angle of the faces concerned must not be changed, and volume of metal must not be materially reduced.

h. A loose barrel, which shakes when assembled, indicates worn barrel head grooves or magazine nut. The gun should be turned over to ordnance personnel for correction or repair.

i. If shell appears unnecessarily loose in chamber with sliding breech locked, the gun should be turned over to ordnance personnel to be checked for headspace, or worn or loose locking block.

j. Adjustment and maintenance of the gun in the case of using troops is limited to the removal and replacement of the parts and groups of parts (pars. 24 and 25), together with cleaning and lubrication, and such adjustments as are necessary in assembling the gun as outlined.

27. CLEANING AND LUBRICATION.

a. Cleaning, oiling, and lubricating may be accomplished in a manner similar to that described for the Winchester M97 Gun (par. 11). Attention should be given to corresponding parts and surfaces when lubricating.

STEVENS SHOTGUN, 12-GAGE, M620A, M520, AND M620

b. The barrel, magazine, and operating handle group should be removed from the receiver for cleaning the bore, and for thorough cleaning the groups should be removed, and the parts and assemblies cleaned, oiled, and lubricated as directed. With the bayonet attachment assembled to the gun, the bore should be cleaned from the muzzle end, and a rag stuffed into the receiver to protect the action while cleaning. When it is necessary to remove the groups, the bore may be cleaned from the breech end (par. 3 i).

c. Points to be lubricated are:
(1) Operating handle bar opening in barrel head and receiver.
(2) Operating handle bar guideway in receiver.
(3) Lifter screw.
(4) Slide guideways in receiver.
(5) Trigger, hammer, and sear pins, and hammer roller (M620 Gun).
(6) Locking block guides.
(7) Slide plunger.
(8) Outer surface of magazine tube where operating handle tube bears.

d. In very cold climates oiling and lubricating should be reduced to a minimum. Only surfaces showing signs of wear should be lightly oiled. Refer to "Special Maintenance," section XI.

SHOTGUNS, ALL TYPES

Section V

ITHACA SHOTGUN, 12-GAGE, M37

	Paragraph
Description	28
Data	29
Operation	30
Functioning	31
Removal of groups	32
Replacement of groups	33
Field inspection	34
Cleaning and lubrication	35

28. DESCRIPTION.

a. Identification marks on this gun are generally to be found as follows:

(1) Name of maker and model number are stamped on top of barrel near the breech end.

(2) The serial number of the gun is stamped on the lower face of the barrel near the breech end and on the lower face of the forward end of the receiver.

b. This gun (figs. 51 and 52) is a manually operated repeating shotgun of the slide action, hammerless, solid-frame, and take-down type. The gun is so constructed that the barrel can be easily removed by disengaging the stud on the magazine nut from the barrel lug, and then unlocking the barrel from the receiver by disengaging interrupted threads on the rear end of the barrel from like threads in the forward end of the receiver. This construction facilitates cleaning and transportation (par. 3 i).

c. This gun is furnished in various grades having barrels of different lengths and degrees of boring, and other modifications of design. Basically, however, the mechanisms of all guns of this make and model are the same. For convenience herein the guns will be classified as three types: riot, sporting skeet, and sporting trap, although variations of these types may occur.

NOTE: The standard grade of this model gun is commercially known as the "Featherlight".

(1) The riot-type gun is usually furnished with a 20-inch plain barrel, bored full cylinder, and may be equipped with a bayonet attachment and hand guard attached to the muzzle end of the barrel, and a leather sling attached to a sling swivel on the bayonet attachment and the stock. If equipped with bayonet attachment it is solid-frame (par. 3 i).

ITHACA SHOTGUN, 12-GAGE, M37

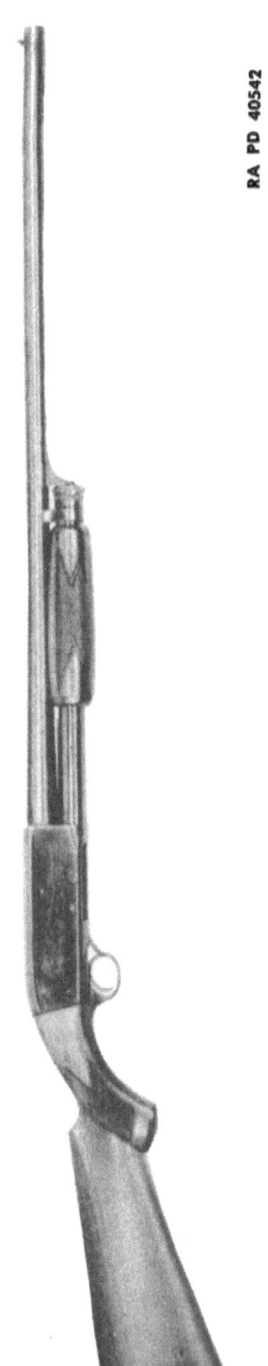

RA PD 40542

Figure 51 — Right Side View — Sporting Trap Type — Ithaca Shotgun M37

RA PD 40543

Figure 52 — Left Side View — Sporting Trap Type — Ithaca Shotgun M37

SHOTGUNS, ALL TYPES

(2) The sporting skeet-type gun is usually furnished with a 26-inch plain or ribbed barrel, bored improved cylinder and is without bayonet attachment or sling.

(3) The sporting trap-type gun (figs. 51 and 52) is similar to the skeet-type gun but is furnished with a 30-inch barrel, bored full choke and is without bayonet attachment or sling.

d. **General Description** (figs. 53, 63 and 64).

(1) The stock of the gun is bolted to the tang on the rear end of the receiver, and the barrel is locked to the forward end by means of interrupted threads on barrel and in receiver, and fastened to the forward end of the magazine by means of a lug on the barrel and a stud on the magazine nut as already explained. The magazine tube is screwed into the receiver at manufacture. The slide handle is mounted and operated on the magazine tube. The rear end of the slide handle bar passes through the forward end of the receiver and engages with and operates the slide, which in turn operates the breechblock, carrier, and shell stops, and cams back and cocks the hammer.

NOTE: The slide handle refers to the slide handle group, which consists of the slide bar, tube, and wooden slide handle assembled to tube.

(2) The receiver contains the operating mechanism and to its lower rear end is attached the trigger plate to which is mounted the firing mechanism and the slide stop. The receiver is open at the bottom to permit loading and the ejection of the fired shell cases. The carrier (figs. 54 and 55) is pivoted to the inside of the receiver and acts to raise the shells to the bore line in loading and ejects them when extracted from the barrel chamber by the extractors on the breechblock. The positive (right) and spring (left) shell stops are pivoted in the forward end of the receiver to the rear of the magazine opening.

(3) The breechblock (figs. 54 and 55) contains the extractors and the firing pin and their components, and mates with and is operated by the slide through the medium of a lug and hooked projection on the bottom of the breechblock seating in mating apertures in the slide. A projection on the rear top of the breechblock seats in front of a locking shoulder cut in the top of the receiver to lock the breechblock in position, when cammed up by the slide in the forward movement.

(4) The slide (figs. 54 and 55) is engaged with the slide bar, as already explained, by means of a spring pin passing laterally through the slide and rear end of the slide bar. The slide operates in guideways in the side walls of the receiver.

(5) The trigger plate contains the trigger, hammer, mainspring, slide stop, and safety together with their components.

ITHACA SHOTGUN, 12-GAGE, M37

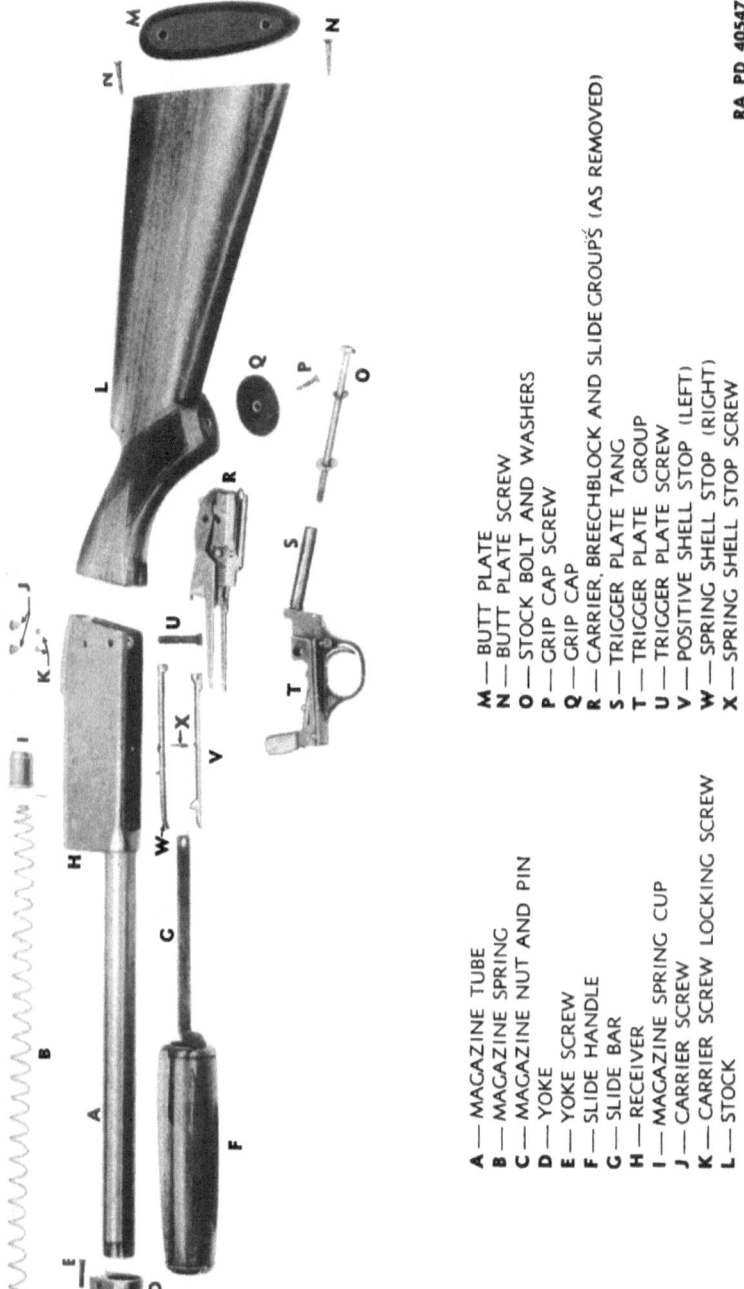

A — MAGAZINE TUBE
B — MAGAZINE SPRING
C — MAGAZINE NUT AND PIN
D — YOKE
E — YOKE SCREW
F — SLIDE HANDLE
G — SLIDE BAR
H — RECEIVER
I — MAGAZINE SPRING CUP
J — CARRIER SCREW
K — CARRIER SCREW LOCKING SCREW
L — STOCK
M — BUTT PLATE
N — BUTT PLATE SCREW
O — STOCK BOLT AND WASHERS
P — GRIP CAP SCREW
Q — GRIP CAP
R — CARRIER, BREECHBLOCK AND SLIDE GROUPS (AS REMOVED)
S — TRIGGER PLATE TANG
T — TRIGGER PLATE GROUP
U — TRIGGER PLATE SCREW
V — POSITIVE SHELL STOP (LEFT)
W — SPRING SHELL STOP (RIGHT)
X — SPRING SHELL STOP SCREW

Figure 53 — Disassembled View (Without Barrel) — Ithaca Shotgun M37

SHOTGUNS, ALL TYPES

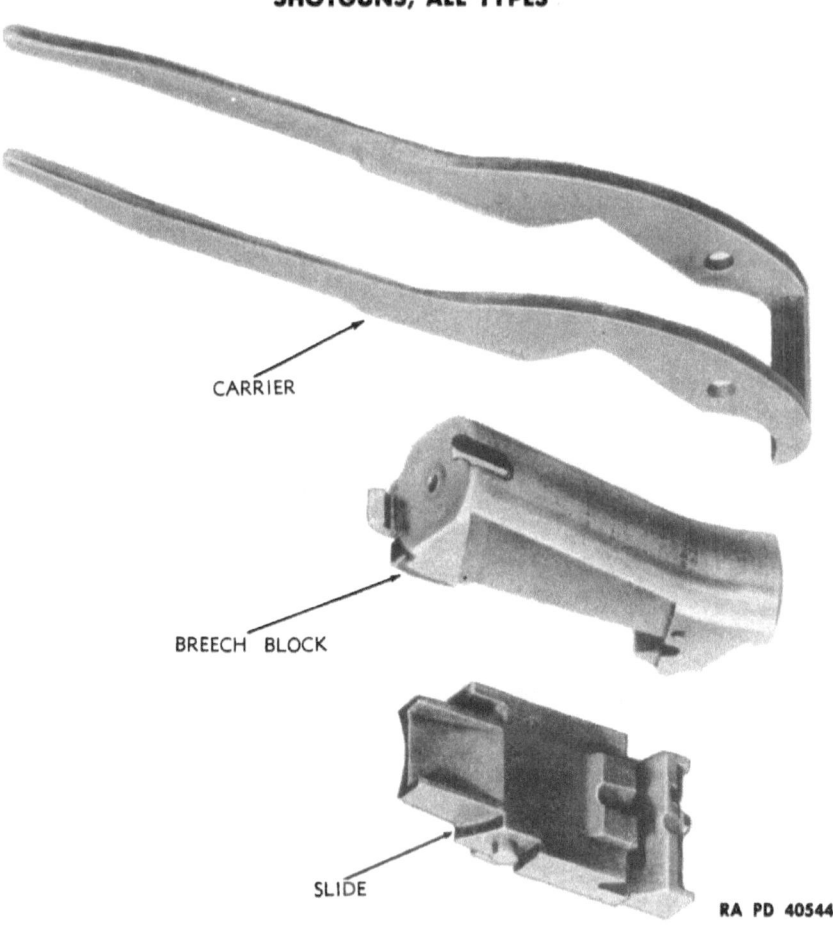

Figure 54—Carrier, Breechblock and Slide—Front, Top, and Left Side View—Ithaca Shotgun M37

(6) The magazine which is of the tubular type is screwed into the receiver below the barrel and has a capacity of four shells loaded end to end. The shells are pressed together and fed into the receiver by the force of the magazine spring acting upon the magazine spring cup.

(7) The positive (right) and spring (left) shell stops are pivoted in the lower part of the receiver and operated by the slide to block and release the shells in the magazine.

(8) The slide stop is pivoted in the trigger plate and acts upon the slide to block its rearward movement after it has locked the breechblock in position, thereby preventing premature unlocking of the breechblock (figs. 59 and 60). The slide stop is disengaged from the slide either by

TM 9-285
28

ITHACA SHOTGUN, 12-GAGE, M37

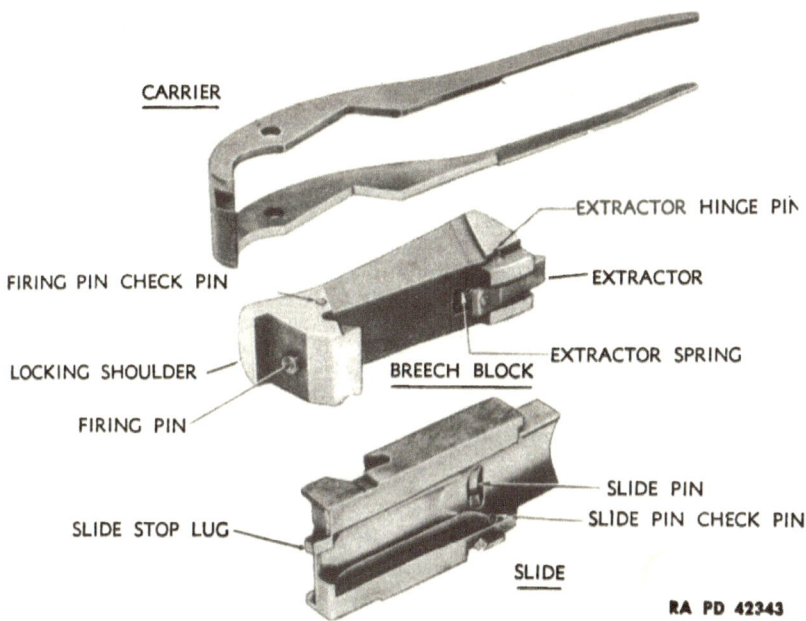

Figure 55—Carrier, Breechblock and Slide—Rear, Bottom and Right Side View—Showing Location of Parts—Ithaca Shotgun M37

the descending hammer or by manual upward pressure on the lower end of the stop which projects through the floor of the trigger plate at the right side of the forward end of the trigger plate guard bow. A hook on the rear end of the slide stop, engaging with a lug on the side of the hammer, locks the hammer back in the extreme rearward position when cammed back by the slide. The rear end of the stop is held in position to engage the hammer by the slide riding over and depressing the forward end of the stop. The forward end of the stop is released and springs up, thereby disengaging the lock from the hammer as the slide reaches the forward position. The hammer is then caught and held by the sear. This prevents premature release of the hammer.

(9) The trigger safety slides laterally in an aperture in the trigger plate guard bow just behind the trigger, and acts to block or clear the trigger thus preventing or allowing retraction to fire the gun.

(10) The bayonet attachment, bayonet, bayonet scabbard, and gun sling (fig. 5) when and if furnished, are similar to those furnished for the Winchester Guns M97 and M12 (par. 4 e (9))

SHOTGUNS, ALL TYPES

29. DATA.

Gage of bore	12
Boring of barrel—riot type	Cylinder
Boring of barrel—sporting skeet type	Improved
Boring of barrel—sporting trap type	Full choke
Type of action	Slide
Type of firing mechanism	Hammerless
Type of magazine	Tubular
Capacity of magazine	4 shells
Length of barrel—riot type	20 in.
Length of barrel—sporting skeet type	26 in.
Length of barrel—sporting trap type	30 in.
Length of stock, receiver, and magazine assembled (approx.)	31¾ in.
Length of assembled gun—riot type (approx.)	40 in.
Length of assembled gun—sporting skeet type (approx.)	46 in.
Length of assembled gun—sporting trap type (approx.)	50 in.
Weight of assembled gun—riot type (approx.)	6 lb
Weight of assembled gun—sporting skeet type (approx.)	6¼ lb
Weight of assembled gun—sporting trap type—"Featherlight" (approx.)	6⅝ lb

30. OPERATION.

a. The gun is operated by moving the slide handle fully and smartly backward and forward. This action unlocks the breechblock, extracts and ejects the fired shell casing, cocks the hammer, transfers a live shell from the magazine to the chamber of the barrel, and relocks the breechblock behind the shell.

CAUTION: During operation, the muzzle of the gun should always be pointing at a safe spot.

b. Before the slide handle can be retracted the slide stop must be disengaged from the slide (fig. 56). If the gun has been fired and the hammer consequently forward, the only movement necessary is to reciprocate the slide handle as above, as the descending hammer has already disengaged the stop. If, however, the hammer is in the cocked position, it is necessary to push up the lower end of the stop which is visible at the right side of the forward end of the trigger plate guard bow, before moving the slide handle to the rear. The slide handle should be moved forward slightly to reduce friction between slide and stop while disengaging.

CAUTION: During these operations, the finger should remain outside the trigger plate guard bow. Reciprocation of the slide handle should be full and smart to insure extraction of the shell, cocking of the hammer, and complete locking of the breechblock, and engagement of the slide

ITHACA SHOTGUN, 12-GAGE, M37

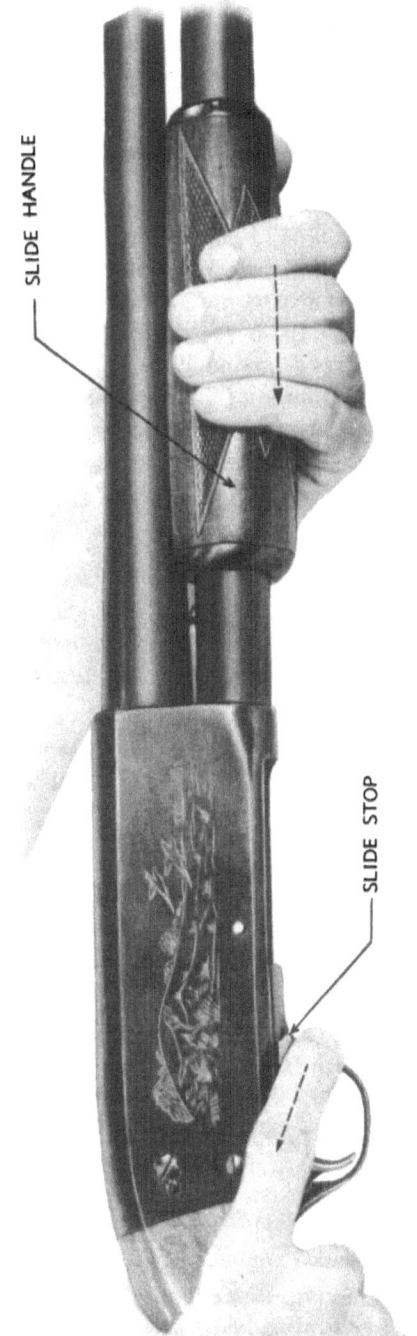

Figure 56—Disengaging Slide Stop from Slide—Ithaca Shotgun M37

SHOTGUNS, ALL TYPES

stop. Slamming of the mechanism, however, should be avoided. When the gun is being fired as a repeater, all pressure should be removed from the trigger while operating.

c. With the gun loaded, locked, and cocked, the only movement necessary to fire the gun is the pulling of the trigger.

d. The trigger safety, positioned in the rear of the trigger plate guard bow, operates laterally (fig. 57). When pushed to the left, the trigger is free to be pulled and the gun fired. When pushed to the right, the trigger is blocked and cannot be pulled nor the gun fired. With a shell in the chamber, it is best to block the trigger by pushing the safety to the right, unless the gun is to be fired immediately.

NOTE: The safety is sometimes furnished for left-handed shooters to operate in opposite direction. This, however, is the exception, but should be checked, with gun empty, on every gun.

e. **To Load and Unload the Magazine.**

(1) To load the magazine (fig. 58), press a shell, nose first, into the rear of the magazine against the magazine spring cup until it slips in front of and is retained by the spring (left) shell stop. Load another shell in the same way, pressing it against the base of the first shell loaded until four in all are loaded. Loading should be done with the breechblock locked.

(2) To unload the magazine, press in on the spring (left) shell stop, with breechblock locked, and allow the shells to slip out one by one.

f. **To Load and Unload the Gun.**

(1) To load a shell from the magazine into the chamber, slide the safety to the safe position. Then disengage the slide stop and reciprocate the slide handle (par. 30 b). Another shell may then be loaded into the magazine. Test locking of the breechblock by attempting to retract the slide handle. The slide handle should not retract. Allow the safety to remain in the safe position unless gun is to be fired immediately.

(2) To unload the gun, slide safety to safe position as above, and unload the magazine as explained in paragraph e (2) above. Then disengage the slide stop and retract the slide handle, to extract and eject the shell in the chamber. Inspect magazine and chamber to make sure gun is fully unloaded.

31. FUNCTIONING.

a. As already explained, the functioning of the operating mechanism is accomplished by the reciprocation of the slide handle. The rear end of the slide bar is engaged with the slide by means of a spring-operated pin running laterally through slide and bar. As the slide bar is moved

TM 9-285
31

ITHACA SHOTGUN, 12-GAGE, M37

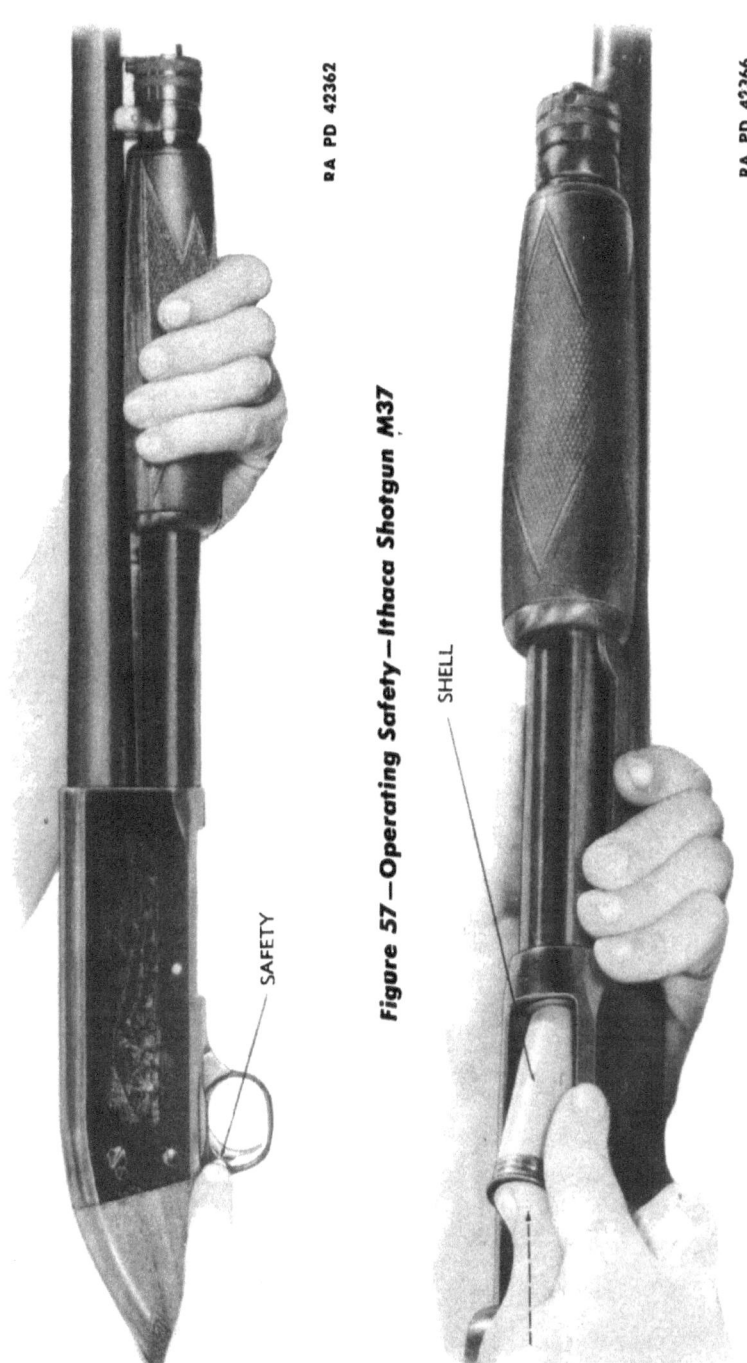

Figure 57 — Operating Safety — Ithaca Shotgun M37

Figure 58 — Loading Magazine — Ithaca Shotgun M37

SHOTGUNS, ALL TYPES

to the rear it moves the slide with it. The slide moves, for a short distance, independent of the breechblock, thus allowing the hooked cam lug on the rear lower face of the breechblock to drop down into a similarly shaped aperture in the top of the slide (fig. 59). This action unlocks the breechblock from the top of the receiver. From this point on, the slide pulls the breechblock with it to the rear. As the slide moves forward again, it pulls the breechblock with it, and at the end of the forward movement cams up the rear end of the breechblock in front of the locking shoulder in the top of the receiver to lock the breechblock to the receiver (fig. 60).

b. During its rearward movement, the slide cams back the hammer, mounted in the trigger plate, and releases a shell from the magazine by cam action on the shell stops. The shell is driven into the receiver by the force of the magazine spring and is retained in the receiver by the carrier which has been cammed down by the slide.

c. When the spring (left) shell stop is cammed back on the rearward movement of the slide so as to release a shell from the magazine, the positive (right) stop is cammed up so as to engage the following shell. This movement is reversed as the slide nears the end of its foward movement, the positive stop is cammed from engagement, and the spring stop engaged. Thus double feeding is prevented.

d. As the slide is pulled forward by the slide bar, it pulls the breechblock with it as already explained. The slide cams up the carrier which lifts the shell in line with the chamber, and the breechblock pushes the shell into the chamber.

e. As the hammer reaches the extreme rearward position, a lug on the hammer is caught and held by a hook on the slide stop, and is so held until the slide nears the forward position and the breechblock has been cammed up in front of the locking shoulder in the top of the receiver (fig. 61). At this point, about ¼ inch from the extreme forward position of the slide, the slide cams the stop from engagement with the hammer and the sear immediately engages and retains the hammer in the cocked position, thus preventing premature firing of the gun.

f. As the slide reaches the extreme forward position, the forward end of the slide stop, over which it has been riding, springs up behind it to block any rearward movement of the slide, thus preventing premature unlocking of the breechblock (fig. 62).

g. The hammer is released to fire the gun by pulling the trigger which levers the sear from engagement with the sear notch in the hammer. The sear is integral with the trigger.

ITHACA SHOTGUN, 12-GAGE, M37

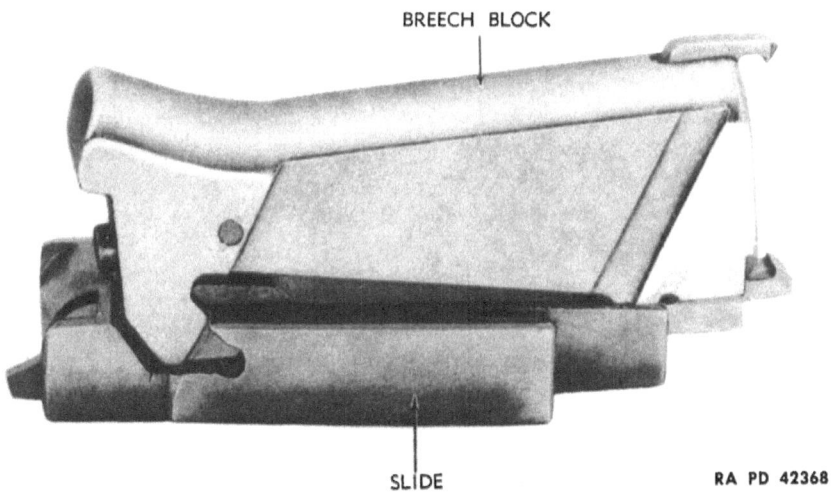

Figure 59 — Breechblock and Slide in Unlocked Position — Right Side View — Ithaca Shotgun M37

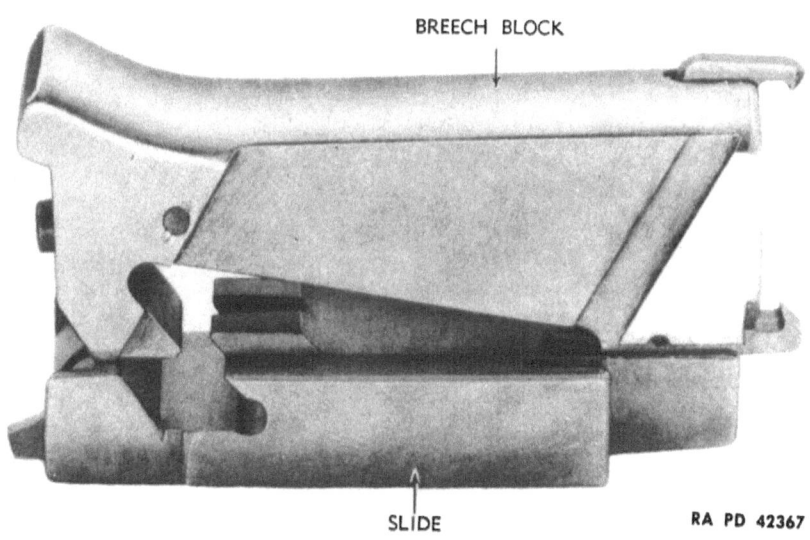

Figure 60 — Breechblock and Slide in Locked Position — Right Side View — Ithaca Shotgun M37

TM 9-285
SHOTGUNS, ALL TYPES

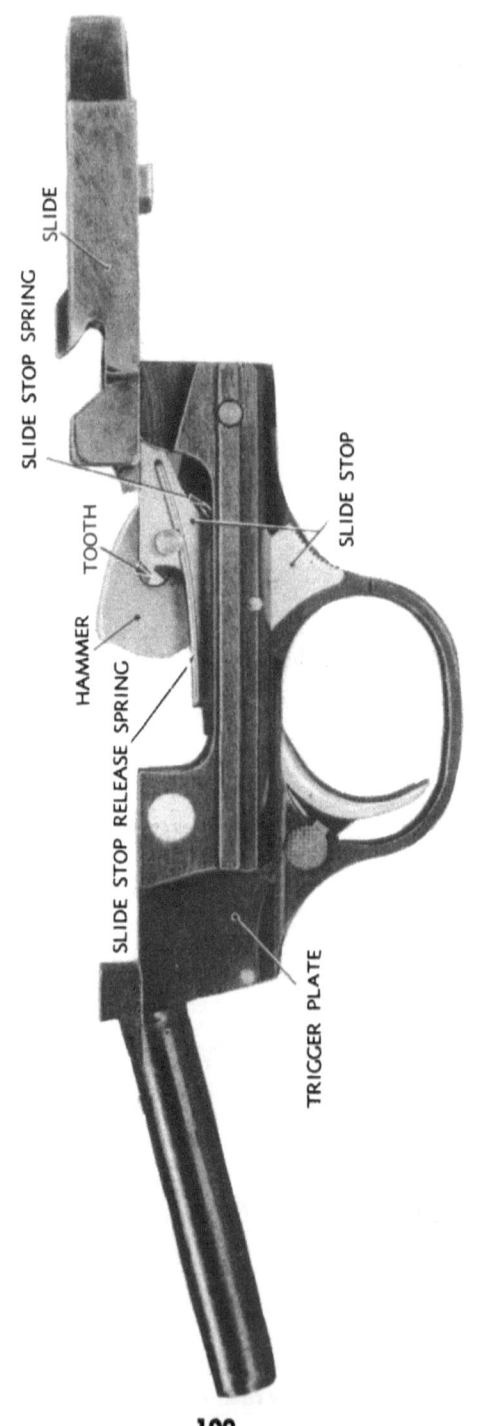

Figure 61 — Slide and Trigger Plate Groups — Slide Riding Over Stop — Hammer Locked — Ithaca Shotgun M37

ITHACA SHOTGUN, 12-GAGE, M37

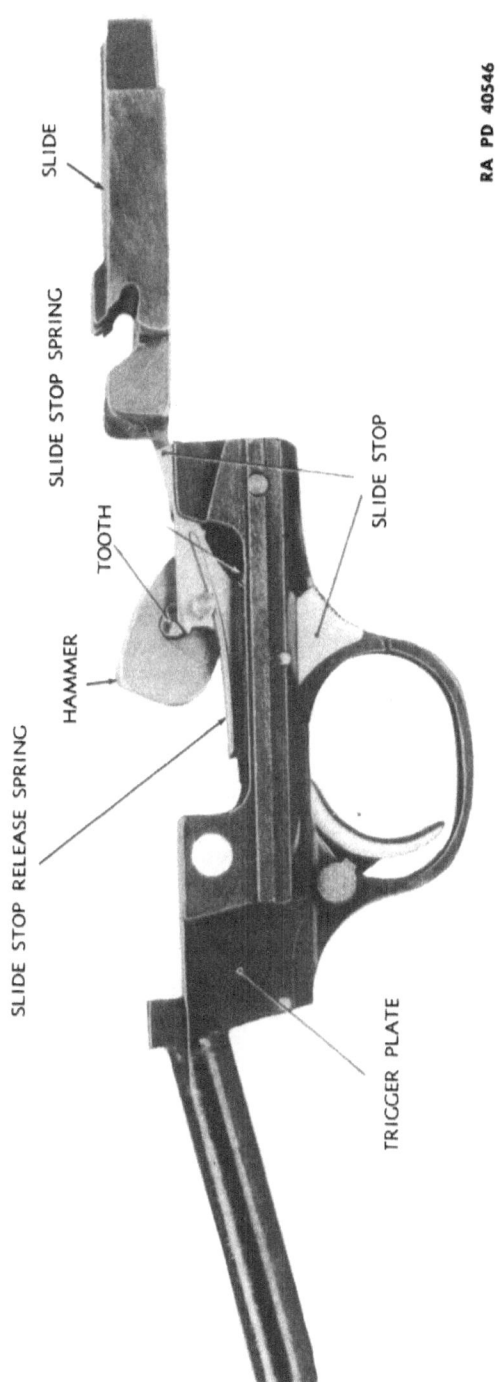

Figure 62 — Slide and Trigger Plate Groups — Slide Blocked — Hammer Unlocked — Ithaca Shotgun M37

SHOTGUNS, ALL TYPES

32. REMOVAL OF GROUPS (figs. 53, 63 and 64).

a. Groups and parts should be removed and replaced in the order given below. Groups and parts when removed should be placed on a clean, flat surface and care observed to prevent loss of screws and small parts. Remove as follows:

(1) BARREL AND BAYONET ATTACHMENT.

(a) The bayonet attachment may be removed in a similar manner to that prescribed for the Winchester Gun M97 (par. 8 a (1)).

(b) To remove the barrel, disengage slide stop, positioned in forward left face of trigger plate guard bow, by pressing upward, and pull the slide handle to the rear. The breechblock must be in the rearward position before the barrel can be removed in order to clear the extractors from the rear of barrel.

(c) Push out the magazine nut pin and using the nut pin as a lever, turn the magazine nut on the forward end of the magazine counterclockwise until the stud on the nut clears the barrel lug with which it is engaged (fig. 65).

(d) Turn barrel ¼ turn clockwise to disengage the interrupted threads on the rear end of the barrel from those in the receiver, and pull the barrel forward out of the receiver (fig. 66). The magazine is screwed into the receiver at manufacture and should not be removed.

(2) TRIGGER PLATE GROUP. To remove the trigger plate group, the stock must first be removed. This can be accomplished by removing the butt plate screws and butt plate and then, using a long shanked screwdriver, unscrew the stock bolt from hole in butt of stock and pull stock to rear from the tang on the receiver. The stock bolt has a square head and if tight, a ⅜-inch socket wrench may be used to remove it. The trigger plate can then be removed as follows:

(a) With the hammer in the forward (fired) position, unscrew trigger plate screw from left rear face of receiver, just above trigger.

(b) Remove screw and pull trigger plate group to the rear out of receiver.

(3) BREECHBLOCK SLIDE AND CARRIER GROUP. These three assemblies should be removed together as follows:

(a) Hold the receiver bottom side up so that magazine faces left; with the fingernail or small flat instrument, pull the slide pin, visible in the forward end of the slide, towards the body, at the same time moving the slide handle forward, until the slide bar is released from the slide.

(b) Unscrew the two carrier screw locking screws, positioned in the right and left walls of the receiver, and then remove the two carrier screws.

TM 9-285
32

ITHACA SHOTGUN, 12-GAGE, M37

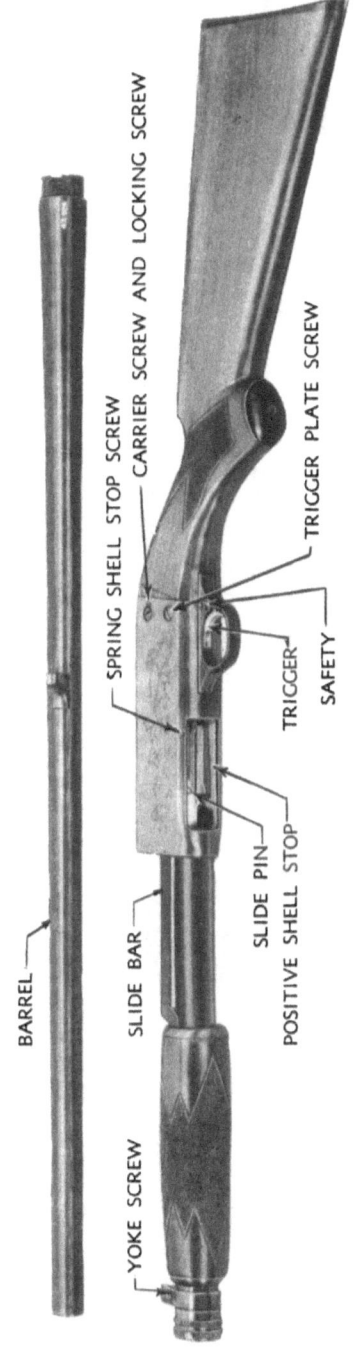

Figure 63 — Gun Taken Down — Left Side View — Showing Location of Parts — Ithaca Shotgun M37

TM 9-285

SHOTGUNS, ALL TYPES

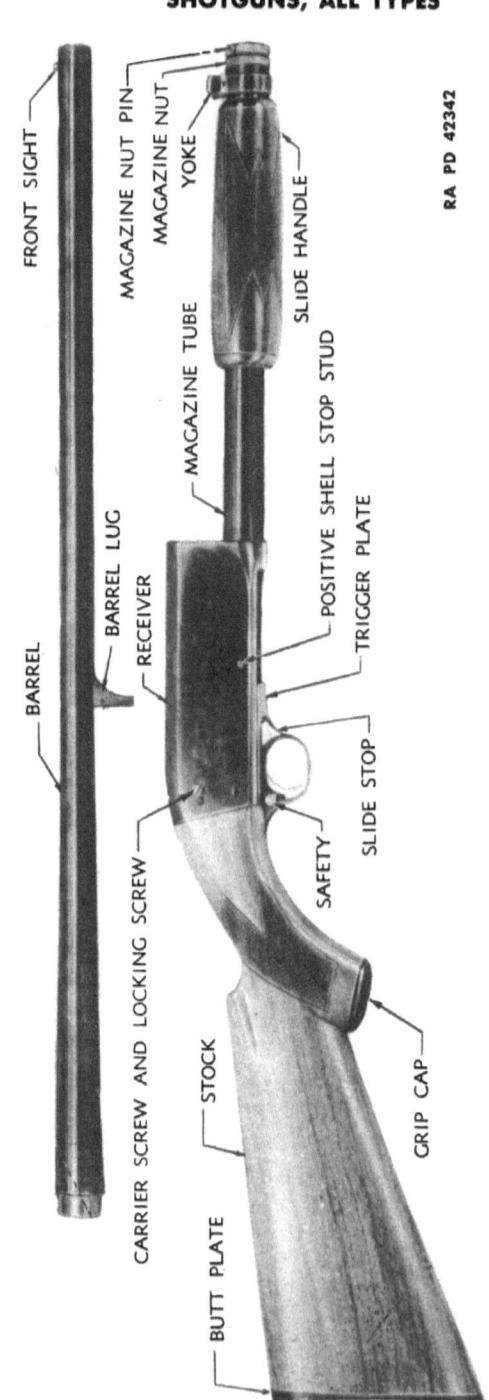

Figure 64—Gun Taken Down—Right Side View—Showing Location of Parts—Ithaca Shotgun M37

ITHACA SHOTGUN, 12-GAGE, M37

Figure 65 — Disengaging Magazine Nut from Barrel Lug — Ithaca Shotgun M37

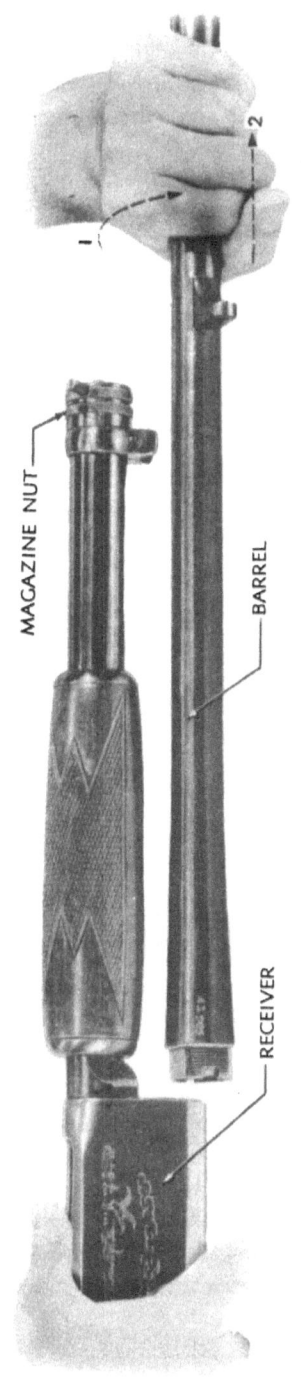

Figure 66 — Barrel Disengaged from Receiver — Magazine Nut in Disengaged Position — Ithaca Shotgun M37

TM 9-285
32-33

SHOTGUNS, ALL TYPES

(c) Pull the carrier, breechblock, and slide together from the rear of the receiver.

NOTE: The positive shell stop should be removed from its groove to prevent loss by pushing inward on pivot stud projecting through right wall of receiver.

33. REPLACEMENT OF GROUPS.

a. Groups and parts should be thoroughly cleaned, lightly oiled, and lubricated, if necessary, before replacing. Replace as follows:

(1) BREECHBLOCK, SLIDE AND CARRIER GROUP.

(a) Replace positive shell stop in its groove in inner right wall of receiver with thin curved end forward. Stop should lie flush in its groove with the pivot stud flush with outer right face of receiver when assembled.

(b) Hold the breechblock bottom side up, with extractor to left, in the left hand. Then, place the slide upon the breechblock so that the hook-shaped aperture in rear of slide mates with hook shaped lug on rear of breechblock.

(c) Taking carrier in right hand with long points to left and curved notched end facing up, slide it on to the breechblock from the rear so that the breechblock lies between the long ends of the carrier with the slide above it, and the small lug on rear end of slide fits in the notch in rear end of carrier.

(d) Transfer group to right hand and take receiver in left hand, bottom side up and open end to right. Then, slide group into receiver through the open end, with the long ends of the carrier pointing into the receiver, so that the guides on either side of the slide mate with the guideways in the inner walls of the receiver; push group into receiver until the slide is checked by rear end of spring (left) shell stop.

(e) Depress forward end of stop and insert small tool under rear end and pry out slightly from receiver wall, until lug on slide will pass under stop. Then, push group all the way forward. Do not pry stop out farther than necessary.

(f) Move carrier to rear until carrier screw holes in carrier and receiver aline; insert carrier screws and screw in snugly until locking screw countersinks in head of carrier screws aline with those in receiver. (Catch threads on both screws before screwing down either screw.) Then, replace locking screws and draw down snugly and flush. Pin end of carrier screw must enter carrier before thread on screw will mate with thread in receiver.

(g) With the slide forward, move the slide handle to the rear until bar contacts slide pin. Retract pin, visible in bottom of slide, with finger-

ITHACA SHOTGUN, 12-GAGE, M37

nail or small flat instrument, and move bar to rear until slide pin engages hole in rear end of bar and release the pin.

(h) Operate slide handle to test operation of slide and locking of breechblock.

(2) TRIGGER PLATE GROUP.

(a) With hammer in forward position and slide forward, push the trigger plate group into lower rear end of receiver, hammer first, so that guides on plate mate with guideways in receiver walls. With hammer in forward position, slide stop will not interfere with slide during replacement of trigger plate group. If hammer is cocked, slide stop must be in disengaged position.

(b) Push plate forward until screw holes in plate and receiver aline and replace trigger plate screw from left side of receiver. Screw in snugly until head of screw is flush with face of receiver.

(c) With trigger plate in position replace the stock by sliding on over tang on receiver until flush with rear end of receiver and trigger plate. Then, replace stock bolt and washer and screw down tightly by turning clockwise. With stock in position, replace butt plate and screw down snugly. If butt plate screws are screwed down too tightly, screw threads in stock may strip out or composition butt plate crack.

(3) BARREL AND BAYONET ATTACHMENT.

(a) Disengage slide stop and pull the slide handle fully to the rear, so that extractor on breechblock will not interfere with barrel during replacement. Inspect bore of barrel for foreign matter. Insert the threaded end of the barrel into the barrel aperture in the forward face of the receiver, with the barrel lug facing to the left (magazine down), and barrel resting in magazine yoke.

(b) Push barrel as far as it will go into receiver so that the interrupted threads on barrel and in receiver are in position to mate, then turn barrel ¼ turn counterclockwise to mate and lock the interrupted threads in barrel and in receiver.

(c) Push out magazine nut pin, and, using as lever, turn nut clockwise to mate stud on forward face of magazine nut with hole in barrel lug. Engage nut snugly with barrel lug, depress pin, and then move slide handle fully forward.

(d) Disengage slide stop, in forward left face of trigger guard bow, by pushing upward, and reciprocate slide handle a few times to test assembly for reciprocation of slide and locking and unlocking of breechblock. Then release stop and push slide handle fully forward to test engagement of stop with slide.

(e) The bayonet attachment may be replaced in a manner similar to that prescribed for the Winchester Gun M97 (par. 9 a(6)).

SHOTGUNS, ALL TYPES

34. FIELD INSPECTION.

a. With the gun completely assembled, test the mechanism for proper functioning. Fired shells may often be used for testing where dummy shells are not available, by turning in the uncrimped end so that the length of the shell will approximate that of a live shell. Use of live shells for testing is prohibited.

CAUTION: Be sure gun is fully unloaded before inspection.

b. Operate the gun as follows:

(1) With the breechblock locked and the hammer cocked, press upward on the lower end of the slide stop, showing at the right side of the forward end of the trigger plate guard bow. Push slide handle forward slightly; then pull smartly and fully to the rear, and then push smartly and fully forward. Reciprocate slide handle thus several times to test smoothness of action.

(2) Retract slide handle as in (1) above, then release slide stop and push slide handle smartly forward to lock the breechblock. Then attempt to retract the slide handle. The slide handle should not retract.

(3) Pull the trigger, thus allowing the hammer to move forward to the fired position and attempt to retract the slide handle. The slide handle should retract.

(4) Retract the slide handle fully and then push forward until the breechblock is fully forward but not raised to the locked position. Then pull the trigger to release the hammer. The trigger should not release the hammer until the breechblock is seated in front of the locking shoulder in the receiver.

NOTE: In this gun the breechblock is seated in the locked position when the slide is still about $\frac{1}{4}$ inch from the extreme forward position and still unblocked by the slide stop. At this point it is possible to release the hammer but it will strike the rear end of the slide and not the firing pin and move forward with the slide.

(5) Place two or more dummy or fired shells in the magazine and work through the action to test gun for feeding, loading, extraction and ejection of shells. The second shell should not leave the magazine until the first shell has been ejected.

NOTE: Fired shells will not work through the action as easily as live or dummy shells as they are somewhat deformed through being fired. Therefore allowance should be made for friction and smoothness of action in positioning the shell.

(6) With breechblock locked and hammer in cocked position, slide trigger safety all the way to the right and attempt to pull the trigger to release the hammer. The trigger should not pull nor the hammer be released.

ITHACA SHOTGUN, 12-GAGE, M37

(7) Slide safety all the way to left and attempt to pull the trigger. The trigger should pull and hammer be released to fire the gun.

CAUTION: A left-hand safety should operate in reverse manner. Guns found to have left-hand safeties should be turned over to ordnance personnel to have safeties replaced with right-hand safeties.

c. When gun does not operate and function smoothly and properly when tested as above, damaged or improperly assembled parts are indicated as follows:

(1) SLIDE HANDLE STICKS. May be due to bent slide bar, bent or broken slide pin, or missing slide pin check pin.

(2) BREECHBLOCK DOES NOT LOCK. May be due to foreign matter on face of breechblock, on locking shoulder in top of receiver or extractor grooves in receiver or rear of barrel, or burs on locking lugs of breechblock, or slide.

(3) HAMMER DOES NOT COCK PROPERLY, OR SLIPS. May be due to burs or foreign matter in sear notch in hammer or worn or burred nose of sear, or weak or broken trigger spring, bent slide stop, or broken hammer bar.

(4) FIRING PIN DOES NOT RETRACT IN BREECHBLOCK. May be due to broken firing pin or spring, or foreign matter in aperture.

(5) SLIDE STOP DOES NOT FUNCTION. May be due to bent slide stop, broken or improperly assembled slide stop springs, burred or worn nose of slide stop or burred or worn stop lug on rear of slide.

NOTE: Slide stop release spring should lie under stud on right side of stop and rear end bear on stop and hammer bar when assembled.

(6) SHELL IS NOT EXTRACTED OR EJECTED. May be due to worn, broken or burred extractors, broken extractor springs, or bent or broken carrier.

(7) TWO SHELLS FED INTO RECEIVER AT ONCE. May be due to broken, bent, or worn positive shell stop or foreign matter in shell stop seating groove in receiver.

(8) SHELL STICKS IN MAGAZINE. May be due to corroded or bent magazine spring cup, dented tube, broken or kinked spring, or foreign matter in tube.

(9) SAFETY STICKS. May be due to burred safety or trigger web, or broken safety catch spring.

d. Inspect barrel and test trigger pull as prescribed in paragraph 3 **n**.

e. In addition to inspection of the gun for operation and functioning, the gun should be inspected generally for condition, and defects noted. Attention should be directed to such defects as cracked wooden parts,

SHOTGUNS, ALL TYPES

cracked or deformed metal parts, dented magazine tube, loose screws or pins, loose or binding parts or assemblies, loose barrel or magazine, loose stock or butt plate, rust, dents, burs or excessive wear of parts. If defects are such that early malfunction of the gun is indicated, the gun should be turned over to ordnance personnel for inspection and correction.

f. Where defects and malfunctions cannot be remedied by cleaning, lubrication, and simple adjustments of assembly, which lie within the scope of using troops, the gun should be turned over to ordnance personnel for a thorough inspection, correction and/or repair.

g. Removal of burs on working parts, trigger adjustments, and like corrections should not be attempted by using troops as stoning of parts must be exacting, the angle of the faces concerned must not be changed, and volume of metal must not be materially reduced.

h. A loose barrel, which shakes when assembled, may be due to improper assembly caused by not inserting barrel far enough into receiver before mating interrupted threads of barrel and receiver, or it may be due to worn parts necessitating adjustment or replacement of barrel or receiver. If due to worn parts, the gun should be turned over to ordnance personnel for correction or repair.

i. If shell appears unnecessarily loose in chamber with breechblock locked, the gun should be turned over to ordnance personnel to be checked for headspace.

j. Adjustment and maintenance of the gun in the case of using troops is limited to the removal and replacement of the parts and groups of parts as outlined in paragraphs 32 and 33, together with cleaning and such adjustments as are necessary in assembling the gun as outlined.

35. CLEANING AND LUBRICATION.

a. Cleaning, oiling, and lubricating may be accomplished in a manner similar to that described for the Winchester Gun M97 (par. 11). Attention should be given to corresponding parts and surfaces when lubricating.

b. The barrel should be removed from the receiver for cleaning the bore and for thorough cleaning, the groups should be removed and the parts and assemblies cleaned, oiled, and lubricated as directed. Gun should be thoroughly cleaned after each two to three thousand shells are fired; oftener if necessary. With bayonet attachment assembled to the gun, the bore should be cleaned from the muzzle end and a rag stuffed into the receiver to protect the action while cleaning. When it is necessary to remove the groups, the bore may be cleaned from the breech end (par. 3 i).

ITHACA SHOTGUN, 12-GAGE, M37

c. Points to be lubricated are:
(1) Slide bar opening in forward end of receiver.
(2) Slide bar guideway in receiver.
(3) Carrier screws.
(4) Slide guideways in receiver.
(5) Mainspring cup.
(6) Trigger and hammer pins.
(7) Outer surface of the magazine tube where slide handle tube bears.

d. In very cold climate, oiling and lubricating should be reduced to a minimum. Only surfaces showing signs of wear should be lightly oiled. Refer to "Special Maintenance," section XI.

SHOTGUNS, ALL TYPES

Section VI

REMINGTON SHOTGUN, 12-GAGE, M10

	Paragraph
Description	36
Data	37
Operation	38
Functioning	39
Removal of groups	40
Replacement of groups	41
Field inspection	42
Cleaning and lubrication	43

36. DESCRIPTION.

a. Identification marks on this gun are generally to be found as follows:

(1) Name of maker on top of barrel near rear end.

(2) Boring of barrel on left side of barrel and serial number of gun on right side near rear end.

(3) The words "REMINGTON (trade mark)," on left side of action slide bar.

(4) The words "REMINGTON (trade mark) model 10," and serial number of gun on left side of receiver.

(5) Serial number of gun on right side of trigger guard tang.

b. This gun (figs. 67 and 68) is a manually operated, repeating shotgun of the slide action, hammerless, solid-frame and take-down type. The take-down gun is so constructed that the barrel, magazine and action bar groups can easily be removed from the receiver by disengaging interrupted threads in rear end of magazine and barrel from like interrupted threads in the receiver, by which the barrel and magazine are locked to the receiver. This construction facilitates cleaning and transportation of the gun (par. 3 i).

c. This gun is usually furnished in the standard grade only, having barrels of different lengths and degrees of boring. Basically, however, the mechanism of all guns of this make and model are the same. For convenience herein the guns will be classified as two types: riot and sporting, although variations of these types may occur.

(1) The riot-type gun (figs. 67 and 68) is usually furnished with a 20-inch plain barrel, bored full cylinder. Some of these guns have a bayonet attachment and hand guard attached to the muzzle end of the

REMINGTON SHOTGUN, 12-GAGE, M10

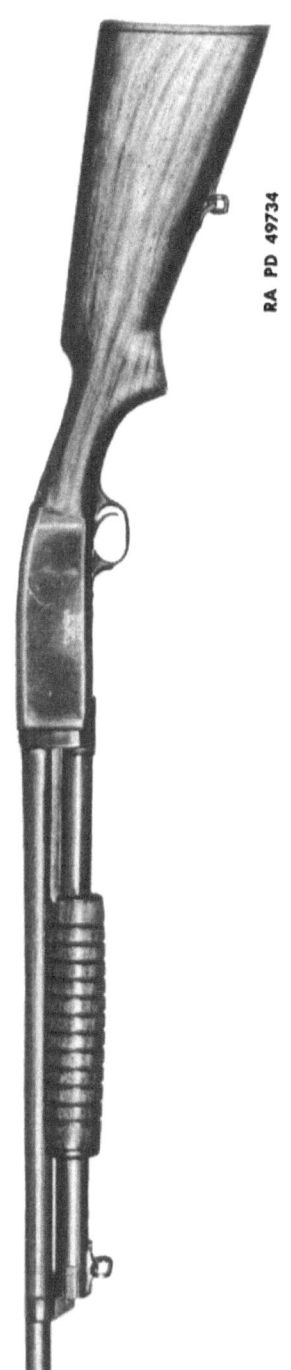

Figure 67 — Left Side View — Riot Type — Remington Shotgun M10

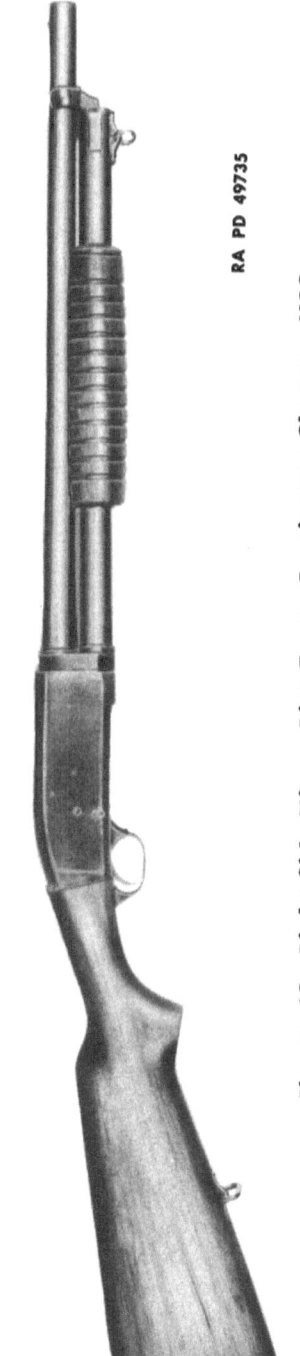

Figure 68 — Right Side View — Riot Type — Remington Shotgun M10

SHOTGUNS, ALL TYPES

barrel and the magazine, and a leather sling attached to a sling swivel on the bayonet magazine tube and the stock. If equipped with bayonet attachment it is solid-frame (par. 3 i).

(2) The sporting-type gun is furnished with a 30-inch barrel, bored full choke, and is without bayonet attachment or sling.

(3) The M10 is no longer made by Remington, but a number are still in use in the field.

d. **General Description** (figs. 69, 70, 77, and 78).

(1) The stock of the gun is fitted to the rear end of the receiver and bolted to the trigger guard tang, and the barrel and magazine locked to the forward end of the receiver as already explained in paragraph b above. The fore end and action bar group is mounted and operates on the magazine tube. The rear end of the action bar passes through the forward end of the receiver and engages with, reciprocates, and cam-operates the breechblock, which in turn engages with and cam-operates the carrier.

(2) The magazine which is of the tubular type is positioned beneath the barrel and has a capacity of five shells loaded end to end. The shells are pressed together and fed into the receiver by the force of the magazine spring acting upon the follower. The magazine is locked to the barrel lug by a lever pivoted in a slot in the magazine plug and held in position when locked by a spring-loaded detent. The magazine spring is held in the tube by the magazine plug and follower.

(3) The receiver contains the operating and firing mechanism with the exception of the trigger, and to its rear end is attached the trigger guard in which is mounted the trigger, trigger spring, and safety slide. The receiver is open at the bottom for loading and ejection of shells and at the rear for passage of the breechblock on its rearward movement. The action bar lock button, which is operated to disengage the action bar lock and allow the action bar to be retracted, is positioned in the right wall of the receiver. The carrier stop which supports the rear end of the carrier is seated in a vertical undercut groove in the rear inner right wall of the receiver. The ejector spring is seated in an undercut groove in the top of the receiver and acts to spring the fired shell down out of the receiver when extracted from the chamber by the extractor on the breechblock. The flap of the carrier assists this function as it is cam-rotated down behind the ejected shell.

(4) The forward end of the carrier (fig. 71) is seated in a blind hole in the right forward, inner face of the receiver and held firm by a spiral friction spring also seated in the hole. The rear end of the carrier pivots on a pin seated in a hole in the carrier stop positioned in the right inner wall of the receiver. The rear end of the carrier is formed into a cam lug

TM 9-285
36

REMINGTON SHOTGUN, 12-GAGE, M10

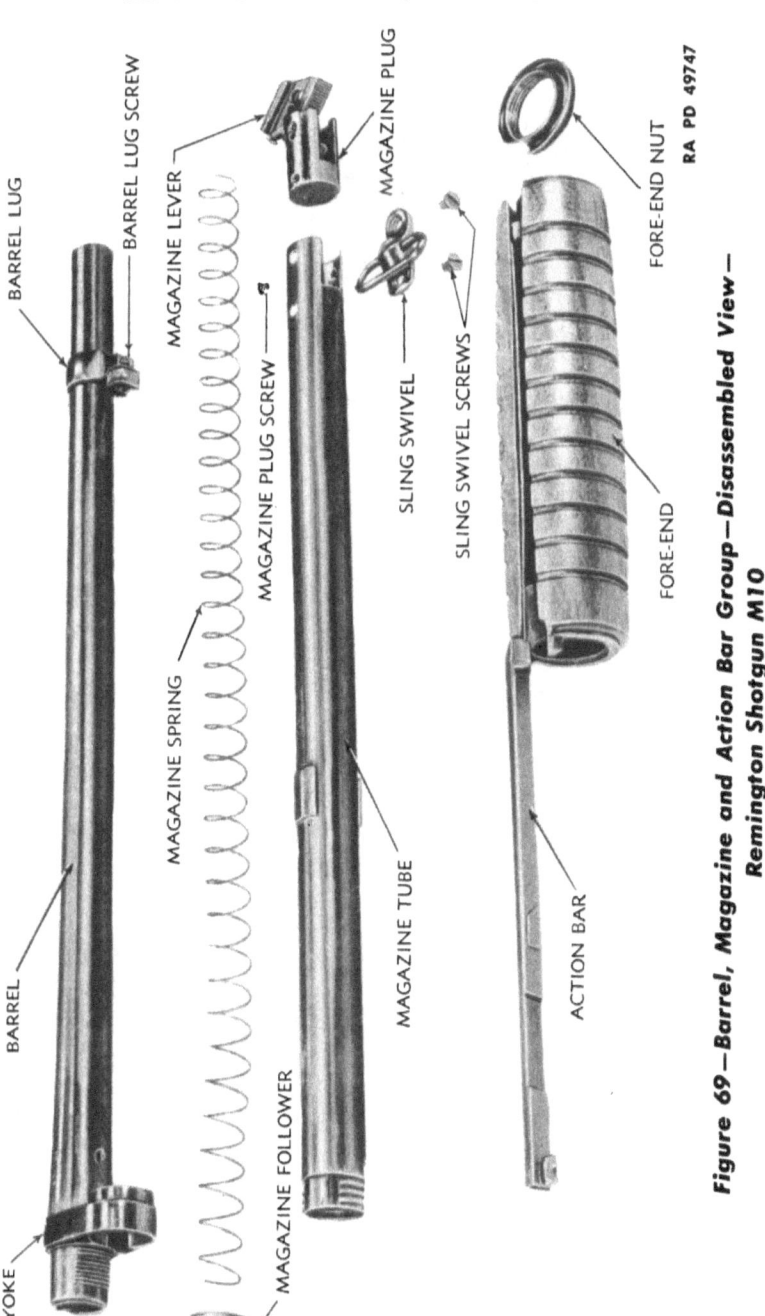

Figure 69 — Barrel, Magazine and Action Bar Group — Disassembled View — Remington Shotgun M10

TM 9-285
36

SHOTGUNS, ALL TYPES

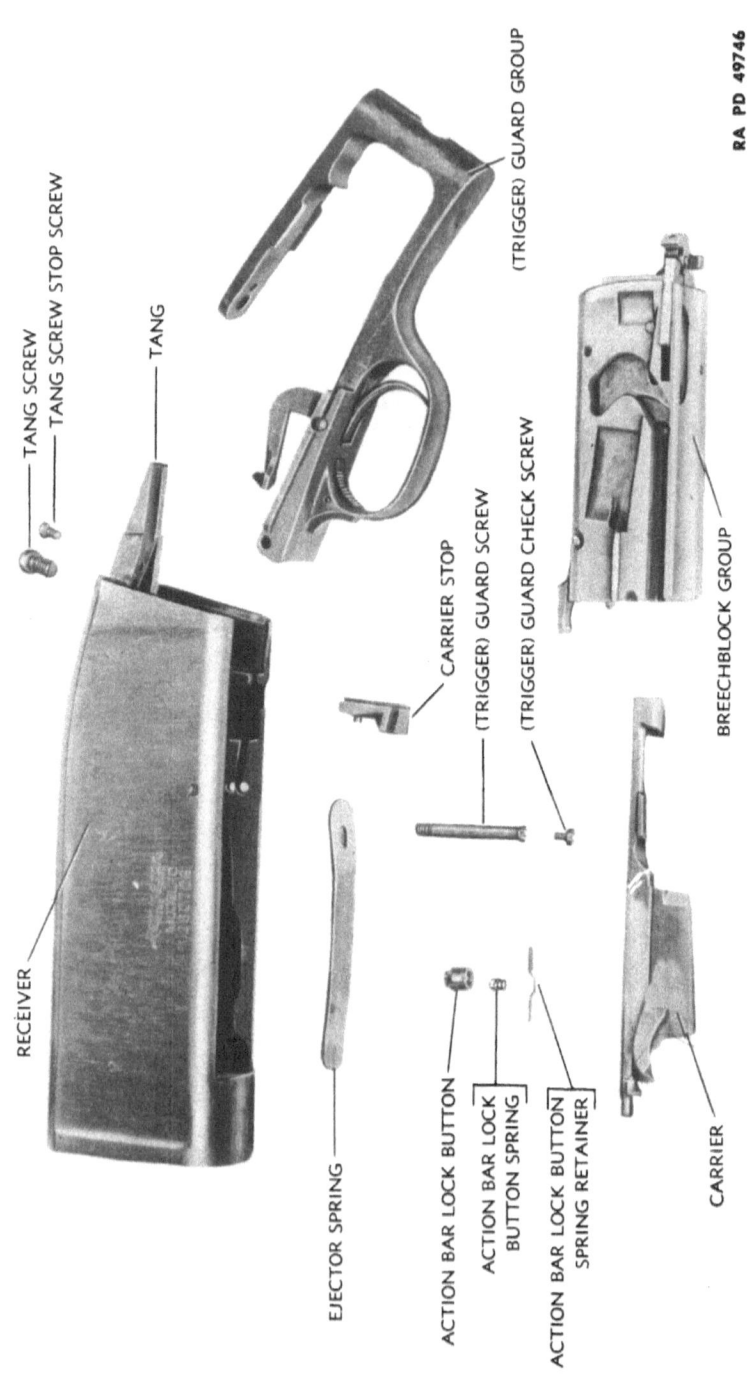

Figure 70—Receiver Group—Disassembled View—Remington Shotgun M10

which seats in a camming aperture in the right forward face of the breechblock and by which the breechblock cam-rotates the carrier in a line parallel to the bore. A curved spring and collar on the forward end of the carrier acts as a shell retainer to block and release the shells from the magazine when the carrier is rotated, and a rectangular flap on the carrier acts to raise the shell to bore line when released from the magazine to enter the receiver. A small triangular cam lug is riveted to the flap to guide the breechblock to engagement with the supporting pins, while passing the flap.

(5) The breechblock (fig. 71) is supported on the left side by the action bar, and on the right side by a guideway in the receiver. The forward end is supported on either side, when in the locked position, by a supporting pin positioned in the receiver wall. When locked, the top of the rear end of the breechblock seats in front of a locking shoulder cut in the top of the receiver.

(6) The firing mechanism with exception of the trigger is mounted in the breechblock and is composed of the firing pin, main spring, firing pin bushing, sear, cocking head, action bar lock, and extractor, together with their springs and components. In some guns of late manufacture, there is a breechblock latch pivoted in the lower rear left face of the breechblock which holds the breechblock in the locked position until disengaged by the rearward movement of the cocking head. In other guns the breechblock is held by the action bar only, which in turn is blocked by the action bar lock. The firing pin is housed in a longitudinal aperture in the breechblock, the rear end passing through the firing pin bushing pinned in the breechblock. The mainspring is compressed between a shoulder on the forward end of the firing pin and the firing pin bushing, and functions the firing pin. The cocking head slides in grooves in the rear end of the breechblock and is connected to, and retracts the firing pin. The cocking head is moved to the rear by a lug on the rear end of the action bar, bearing upon its forward face, as the action bar is moved rearward. The action bar unlocks, reciprocates and locks the breechblock by means of the lug referred to above, which seats in a camming groove in the left side of the breechblock. When moved rearward, the cocking head is caught and retained by the sear, also mounted in the rear end of the breechblock until the sear is levered from engagement by retraction of the trigger.

(7) The action bar lock is pivoted in the rear end of the breechblock. The L-shaped forward end passes laterally through the breechblock, and a spring-loaded shoe on the rear end bears upon the cocking head when in the rearward position. The lock is put under spring tension by bearing upon the cocking head when in the rearward (cocked) position, and the right wall of the receiver. When the breechblock is locked in the forward

SHOTGUNS, ALL TYPES

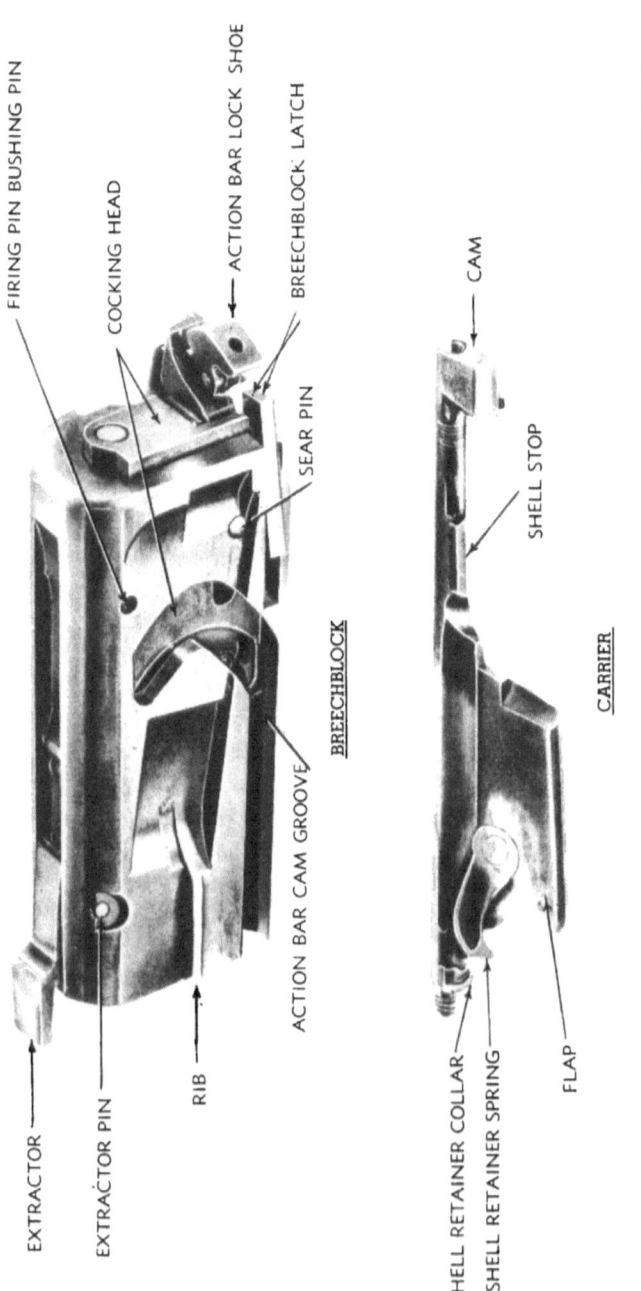

Figure 71—Breechblock and Carrier Groups—Rear, Top and Left Side View—Showing Location of Parts—Remington Shotgun M10

REMINGTON SHOTGUN, 12-GAGE, M10

position, the forward end of the lock springs into a notch in the action bar cam lug and thus prevents retraction of the bar until released. The lock is disengaged either by the forward movement of the cocking head releasing the spring pressure, or by manual pressure on the spring-loaded lock button extending through the right wall of the receiver.

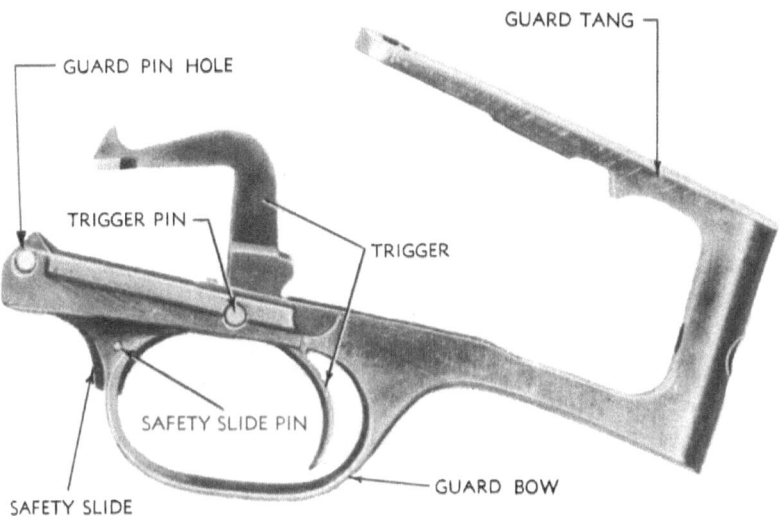

Figure 72 — Trigger Guard Group — Left Side View — Remington Shotgun M10

(8) The extractor is of the claw type, positioned in the top of the forward end of the breechblock and actuated by a leaf spring.

(9) The safety slide is positioned in the forward end of the trigger guard bow (fig. 72) and operates longitudinally to block and release the trigger for retraction.

37. DATA.

Gage of bore	12
Boring of barrel—riot type	Cylinder
Boring of barrel—sporting trap type	Full choke
Type of action	Slide
Type of firing mechanism	Hammerless
Type of magazine	Tubular
Capacity of magazine	5 shells
Length of barrel—riot type	20 in.

SHOTGUNS, ALL TYPES

Length of barrel—sporting trap type . . . . 30 in.
Length of stock and receiver (approx.) . . . . 20 in.
Length of assembled gun—riot type (approx.) . . . . 39½ in.
Length of assembled gun—sporting trap type (approx.) . . . . . . 49½ in.
Weight of assembled gun—riot type—without bayonet attachment (approx.) . . . . 7½ lb
Weight of assembled gun—sporting trap type (approx.) . . . . lb

38. OPERATION.

a. The gun is operated by moving the fore end smartly and fully backward and forward. This action unlocks the breechblock, extracts and ejects the fired shell, cocks the cocking head which is attached to the firing pin, transfers a live shell from the magazine to the chamber of the barrel, and relocks the breechblock behind the shell.

CAUTION: During operation, the muzzle of the gun should always be pointed at a safe spot.

b. Before the action slide can be retracted, the action bar lock must be disengaged from the action bar. If the gun has been fired and the cocking head consequently in the forward position, the only movement necessary is to reciprocate the fore end as above, as the forward movement of the cocking head and subsequent spring action of the action bar lock button has disengaged the action bar lock from the action bar. If, however, the cocking head is in the rearward (cocked) position, it is necessary to depress the action bar lock button protruding through the right wall of the receiver to disengage the lock before retracting the fore end (fig. 73).

CAUTION: During these operations, the finger should remain outside the trigger guard bow. Reciprocation of the fore end should be full and smart to insure extraction of the shell, cocking of the cocking head, and complete locking of the breechblock and engagement of the action bar lock. Slamming of the mechanism, however, should be avoided. When the gun is being fired as a repeater, all pressure should be removed from the trigger while operating.

c. With the gun loaded, locked, and cocked the only movement necessary to fire the gun is to pull the trigger.

d. The trigger safety slide positioned in the forward end of the trigger guard bow operates longitudinally (figs. 75 and 76). When pushed to the rear, the trigger is blocked and cannot be pulled to fire the gun. When pushed forward, the trigger is free to be pulled and the gun fired. With a shell in the chamber, it is best to block the trigger by pushing the safety slide to the rear, unless the gun is to be fired immediately. In this gun the safety slide is easily pushed forward with the finger inside the trigger guard bow when the gun is mounted to the shoulder to fire.

TM 9-285
38

REMINGTON SHOTGUN, 12-GAGE, M10

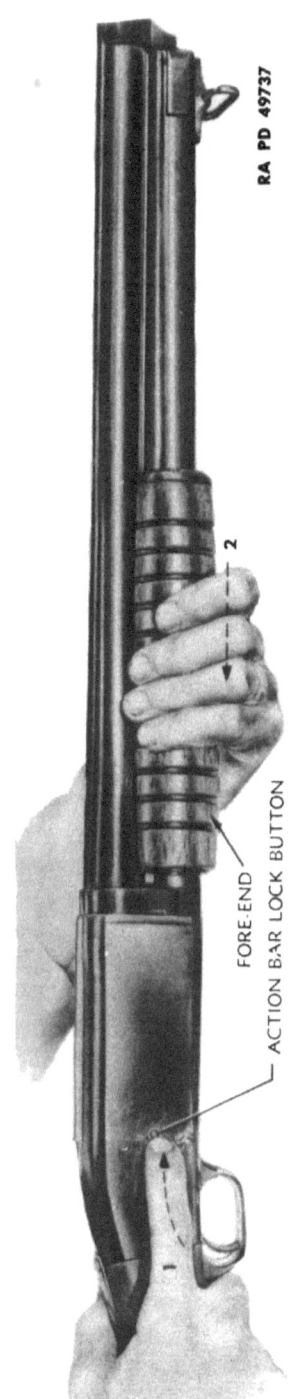

Figure 73 — Retracting Fore End (Action Bar) — Action Bar Lock Button Depressed — Remington Shotgun M10

Figure 74 — Loading Magazine — Remington Shotgun M10

SHOTGUNS, ALL TYPES

e. **To Load and Unload the Magazine.**

(1) To load the magazine (fig. 74), press a shell, nose first, into the rear of the magazine against the magazine follower until it slips in front of and is retained by the curved spring on the forward end of the carrier. Load another shell in the same way until five in all are loaded. Loading should be done with the breechblock locked.

(2) To unload the magazine, push the safety slide to the rear to lock the trigger, depress the action bar lock button to release the action bar, and reciprocate the fore end until all the shells in the magazine are ejected. Then, with breechblock retracted, inspect chamber and magazine to make sure the gun is fully unloaded.

f. **To Load and Unload the Gun.**

(1) To load a shell from the magazine into the chamber of the barrel, push the safety slide to the rear to block the trigger, depress the action bar lock button to disengage the lock from the action bar, and reciprocate the fore end. Another shell may then be loaded into the magazine. Test breechblock for locking by attempting to retract the fore end. The fore end should not retract.

(2) To unload the gun, proceed as in unloading the magazine as explained in e (2) above.

g. **To Load the Chamber Only.** To load the chamber only, push the safety slide to the rear to lock the trigger. Load one shell into the magazine and then reciprocate the fore end as explained above. Another method is to retract the fore end, and with breechblock thus in the rearward position, insert a shell horizontally with nose forward, up into the receiver through the ejection opening. Then push shell forward into the chamber until retained by the ejector spring, and lock the breechblock by pushing the fore end all the way forward. Test locking of breechblock as in f (1) above.

39. FUNCTIONING.

a. As already explained, the functioning of the operating mechanism is accomplished by the reciprocation of the action bar by means of the fore end. A cam lug on the rear end of the action bar seats in an irregular camming aperture in the left side of the breechblock and as the bar is removed rearward, the rear end of the breechblock is cammed down from in front of the locking shoulder in the top of the receiver and the breechblock moved to the rear. As the action bar is moved forward by means of the fore end, the breechblock is moved forward and the rear end cammed up in front of the locking shoulder, at the end of the forward movement.

b. As the action bar moves to the rear in the camming groove of

REMINGTON SHOTGUN, 12-GAGE, M10

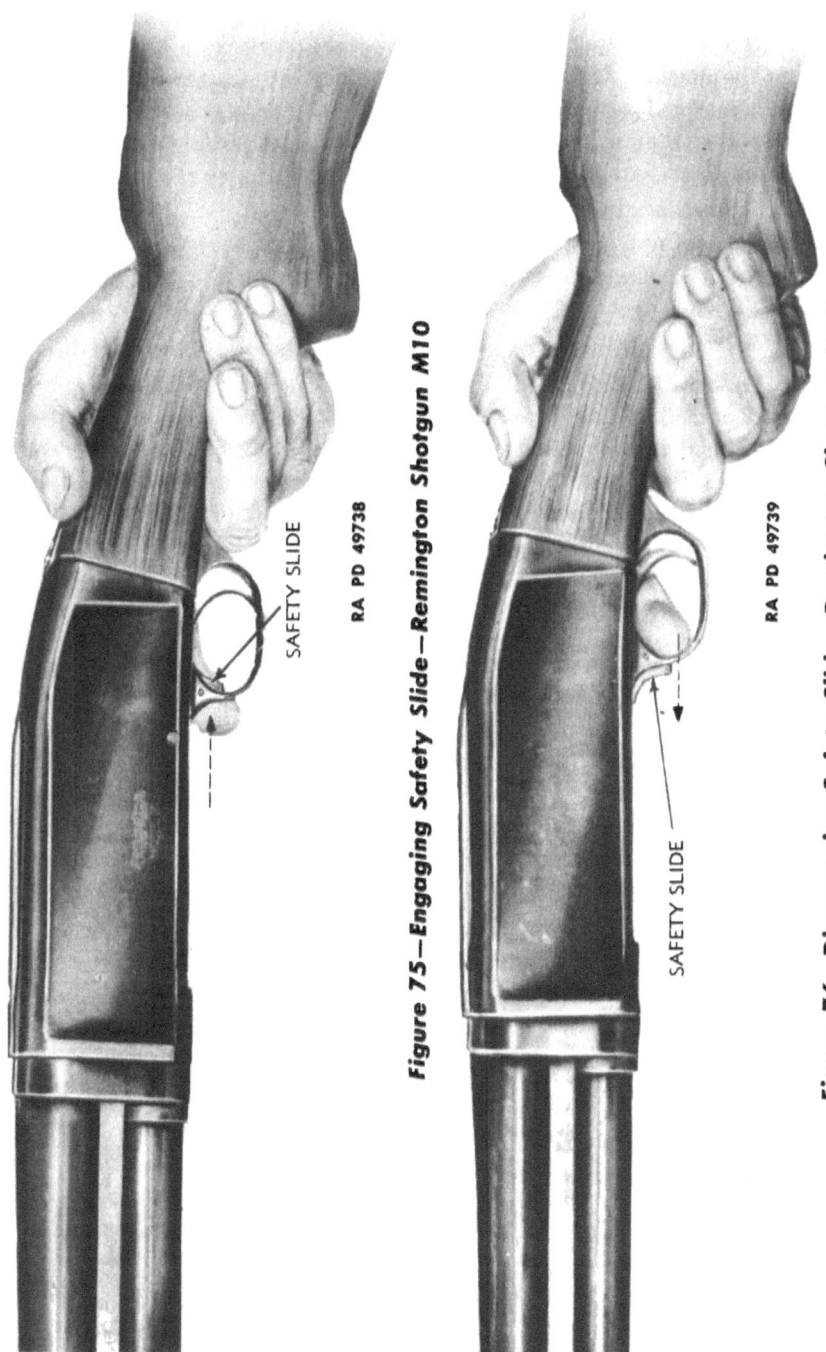

Figure 75—Engaging Safety Slide—Remington Shotgun M10

Figure 76—Disengaging Safety Slide—Remington Shotgun M10

SHOTGUNS, ALL TYPES

the breechblock, the lug on the rear end of the bar strikes the cocking head, positioned in the rear end of the breechblock, and pushes it to the rear against the force of the main spring until caught and held by the sear which is also positioned in the rear end of the breechblock. The cocking head is attached to the firing pin and thus pulls it to the rear. The action bar then continues to push the breechblock to the rear to the end of the rearward movement.

c. The carrier is positioned longitudinally in the right side of the receiver. The forward end seated in a blind hole in the forward end of the receiver, and the rear end in a seat in the carrier stop as already explained. The cam on the extreme rear end of the carrier seats in a camming groove in the right side of the breechblock. As the breechblock is moved backward and forward, it cam-rotates the carrier so that the flap on the carrier moves up and down in the receiver. As the breechblock nears the end of the rearward movement, it rotates the flap of the carrier down to a position below the rear end of the magazine. As the carrier rotates down, the curved shell retaining spring on the forward end, which has been holding the rearmost shell in the magazine, moves from engagement with the base of the shell and the shell thus released is forced out of the magazine into the receiver by the force of the magazine spring. As the retaining spring rotates from engagement with the rearmost shell, a small collar on the carrier, just forward of the spring, rotates down into position to intercept the following shell, thus preventing double feeding. The released shell moves rearward until its base strikes the shell stop lug located on the carrier shank to the rear of the flap. This shell stop lug prevents the shell from striking the breechblock. As the breechblock moves forward, it rotates the carrier upward and the carrier lifts the shell to bore line in time for the breechblock to push it forward into the chamber of the barrel. As the carrier rotates upward, the retaining spring on the forward end rotates into position to again block the rearmost shell in the magazine, thereby taking over the function of the retaining collar which has been holding the shell while the spring was disengaged from it.

d. The ejector spring is positioned longitudinally in the top of the receiver to the rear of the chamber of the barrel when assembled. As the fed shell is lifted to bore line by the carrier, it is pressed against the ejector spring, and as the breechblock moves forward to chamber the shell, it in turn presses upon and keeps the spring pressed upward under tension. As the fired shell is pulled from the chamber by the extractor on the breechblock when the breechblock is moved rearward, it again comes under spring pressure of the ejector spring. As soon as the nose of the shell clears the chamber, the shell is sprung downward out of the receiver by the spring force of the ejector spring bearing upon it. The

REMINGTON SHOTGUN, 12-GAGE, M10

carrier is immediately rotated downward to assist the ejection and clear the receiver, as the breechblock reaches the end of the rearward movement.

e. The action bar lock is pivoted in the rear end of the breechblock with the long L-shaped forward end of the lock passing laterally through the breechblock from right to left. A spring-loaded hinged shoe, pivoted on the rear end of the lock, bears upon the cocking head when in the rearward (cocked) position. The shoe bearing thus, levers the rear end of the lock towards the right wall of the receiver and the L-shaped nose thus is pivoted towards the left wall, through the breechblock, and into the notch in the action bar lug as the breechblock reaches the locked position. Thus the bar is locked in position and cannot be retracted to unlock the breechblock until the bar lock is disengaged. This is accomplished, as explained, either by pressing in on the lock button which extends through the right wall of the receiver or by the forward movement of the cocking head when the gun is fired. The lock button bears upon the rear end of the lock when the breechblock is in the locked position and is under spring tension inward, but the spring force of the lock shoe, when pressing upon the cocking head, is greater than that of the button and thus holds the lock in place despite the spring force exerted by the button. When the lock button is manually pressed inward with a force greater than the opposing force of the lock shoe, it pivots the forward L-shaped end of the lock from engagement with the action bar lug. Likewise, when the cocking head moves forward, the pressure on the lock shoe is released, and the forward end of the lock is levered out of the slot in the action bar lug by the spring force of the lock button bearing upon the rear end of the lock, as the spring force exerted by the shoe, while bearing on the cocking head, no longer opposes the spring force of the button. The action bar is thus freed to unlock the breechblock and move it to the rear when the fore end is retracted. This locking of the action bar prevents premature unlocking of the breechblock.

f. The safety slide operates longitudinally in the forward end of the trigger guard bow, and is held in position by a spring-loaded steel ball, positioned in the guard bow, seating in notches in the slide. When the slide is in the rearward position, it blocks the forward end of the trigger from pivoting downward and thus the trigger from being pulled to lever the sear from engagement with the cocking head. When the slide is in the forward position, it no longer blocks the trigger and it can be pulled to act upon the sear to release the cocking head and thus fire the gun.

g. Some guns of late manufacture have a breechblock latch, pivoted in the lower left, rear face of the breechblock. The rear end of the latch is forced out by the cocking head, when in the rearward position; the forward end is thus forced inward and in this position, the latch is inop-

SHOTGUNS, ALL TYPES

erative. When the cocking head moves forward, it releases the rear end of the latch, and the forward end then springs out into the action bar groove in the receiver and prevents the rear end of the breechblock from being moved downward from in front of the locking shoulder in the top of the receiver. When the cocking head is again moved to the rear by the action bar, it cams the rear end of the latch outward and thus rotates the forward end inward and out of the action bar groove in the receiver. The rear end of the breechblock is thus freed to be moved down and unlocked by the action bar. The latch prevents an unlocked breechblock in case of a hangfire.

40. REMOVAL OF GROUPS (figs. 69, 70, 77, and 78).

a. Groups and parts should be removed and replaced in the order given below. Groups and parts, when removed, should be placed on a clean, flat surface and care taken to prevent loss of screws and small parts. Remove as follows:

(1) BARREL, MAGAZINE AND ACTION BAR GROUP (TAKE-DOWN GUN).

(a) With the breechblock locked, press magazine lever detent, positioned in forward end of magazine lever, down away from the barrel. At the same time push in on the short knurled surface of the detent to swing the rear end of the lever out of the magazine plug, to a right angle with the magazine (tube fig. 79).

(b) Then, using lever, turn the magazine ¼ turn clockwise to disengage it from interrupted threads in receiver, and move it forward as far as it will go (fig. 80).

(c) Grasp the fore end and move the action bar forward until it clears the receiver (fig. 81), then turn the barrel ¼ turn clockwise, to disengage it from interrupted threads in receiver, and group pull from the receiver (fig. 82).

(2) BUTT STOCK.

(a) Remove the butt plate by unscrewing the two butt plate screws and lifting butt plate from butt stock. The butt stock must be removed from the receiver before the breechblock and carrier can be removed.

(b) With a long-shanked screwdriver inserted in the stock bolt aperture now visible in butt stock, unscrew the butt stock bolt by turning counterclockwise until free from the trigger guard tang.

(c) Pull the butt stock to the rear from the trigger guard and receiver.

(3) TRIGGER GUARD GROUP.

(a) With breechblock forward, unscrew tang screw check screw from rear top of receiver and then remove the tang screw from the receiver.

REMINGTON SHOTGUN, 12-GAGE, M10

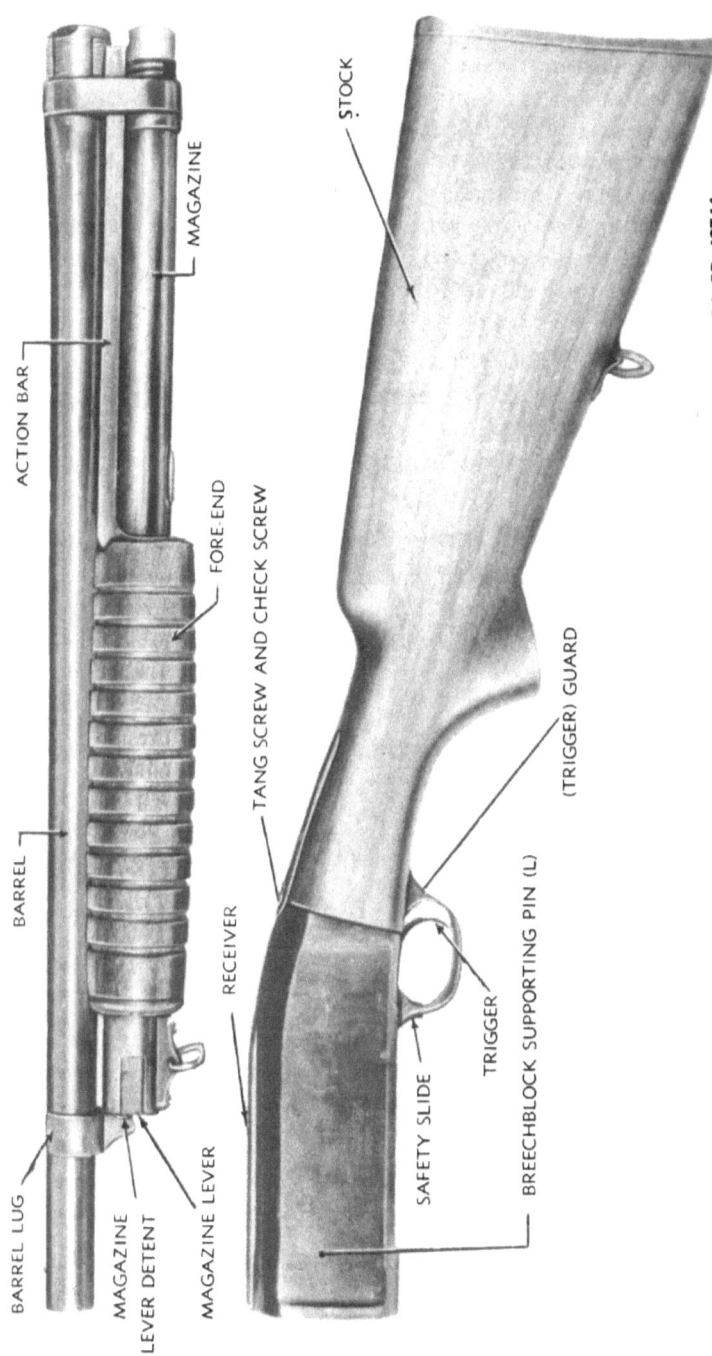

Figure 77 — Gun Taken Down — Left Side View — Showing Location of Parts — Remington Shotgun M10

TM 9-285
40

SHOTGUNS, ALL TYPES

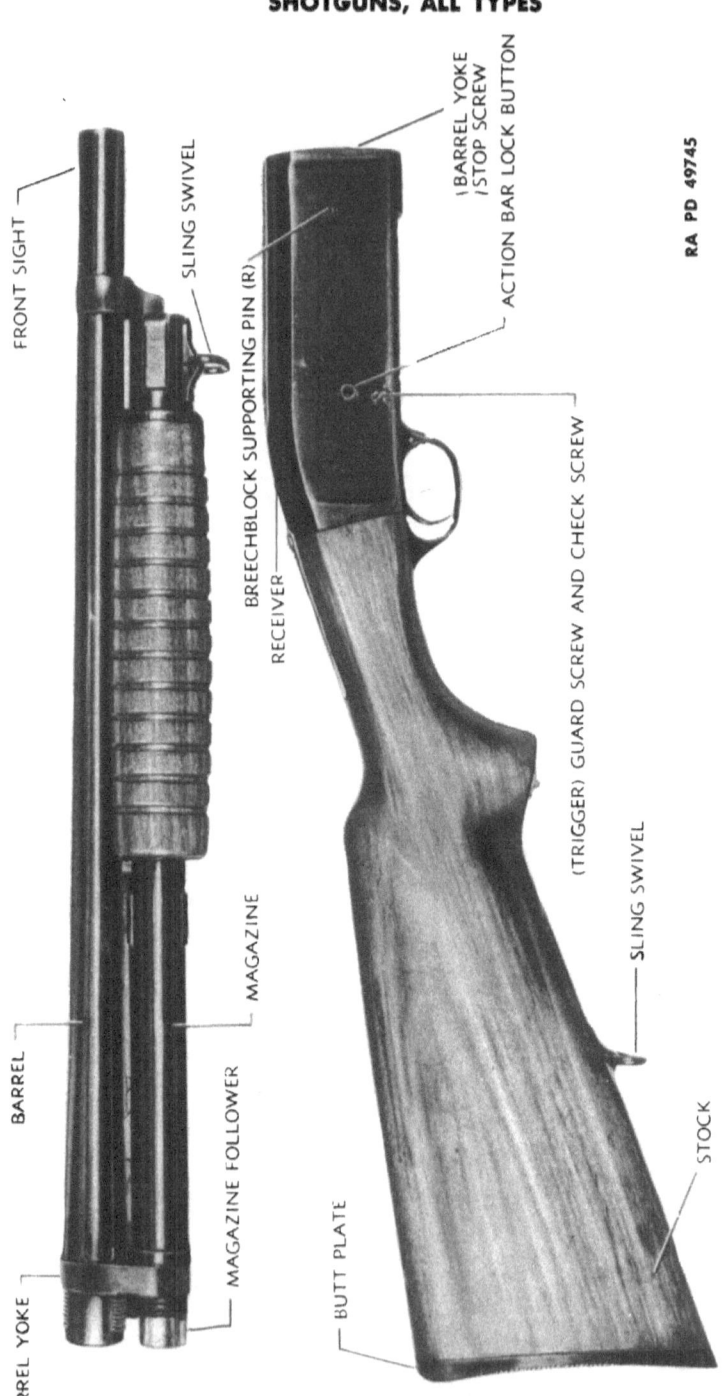

Figure 78—Gun Taken Down—Right Side View—Showing Location of Parts—Remington Shotgun M10

REMINGTON SHOTGUN, 12-GAGE, M10

(b) Unscrew the guard screw check screw from rear right face of receiver and remove the guard screw from the receiver.

(c) Slide the guard to the rear from the receiver.

(4) BREECHBLOCK AND CARRIER GROUP.

(a) Turn the receiver so that the bottom opening is uppermost and the tang of the receiver points to the right. The receiver will be considered as resting in this position until the group is removed.

(b) Press carrier slightly forward (left) and slide the carrier stop upward and remove it from its T-slot in the right inner, rear wall of the receiver. This releases the rear end of the carrier assembly.

(c) Depress the action bar lock button projecting from the right side of the receiver to release the action bar, and with the button depressed, insert the blade of a screwdriver beneath the rear end of the breechblock and pry up the breechblock firmly (do not force) until it starts to rise, then move it with the fingers.

NOTE: If the cocking head is forward, pry up on the breechblock without pressing action bar lock button as action bar lock will be disengaged. However, in guns which have the breechblock latch assembled to the lower rear, left side of the breechblock, the latch must be released by prying the rear end of the latch towards the receiver before the breechblock can be unlocked. If cocking head is to rear (cocked position), this is not necessary but in this case the lock button must be pressed as above.

(d) As the rear end of the breechblock is raised clear of the locking shoulder in the top of the receiver, start moving it to the rear of the receiver by pushing on the face of the breechblock through the barrel opening in the receiver. Press action bar lock shoe, at rear end of breechblock, towards the right wall of receiver to clear lock from carrier stop groove in right wall of receiver as breechblock starts out of receiver. When the breechblock has moved to the rear sufficiently to cam the carrier flap to its horizontal position, grasp the carrier and pull to the rear from its seat in the forward end of the receiver. Then, grasp breechblock and carrier together and pull to the rear out of the receiver, rotating the carrier slightly downward (towards top of receiver) in progress. Do not force. The carrier friction spring seated in the carrier seat in the receiver should be removed to prevent loss, by pulling out of seat. The action bar lock button in right side of receiver and ejector spring in top of receiver should not be removed. Exercise care not to bend ejector spring as proper feeding and ejection of the shell depends upon the flexibility and position of the spring.

41. REPLACEMENT OF GROUPS.

a. Groups and parts should be thoroughly cleaned, lightly oiled, and

SHOTGUNS, ALL TYPES

lubricated, if necessary, before replacing. Replace as follows:

(1) BREECHBLOCK AND CARRIER GROUP.

(a) Before the breechblock and carrier are replaced in the receiver, the carrier friction spring should be replaced in its blind hole in right forward, inside face of the receiver and the ejector spring checked to see that it is all the way to the rear in its seat.

(b) Turn the receiver bottom side up with the barrel opening to the left. Then, place the carrier in its groove in the right wall of the receiver with cam to rear and flap roughly horizontal. Allow the carrier cam to project about an inch outside the receiver, and hold with the left hand.

(c) Grasp breechblock so that extractor is down and pointed to left and mate the carrier cam with the camming aperture in the right face of the breechblock, keeping the flap of the carrier roughly horizontal.

(d) Start the breechblock slowly into its guide grooves in the receiver and move both carrier and breechblock slowly forward (to left), with the carrier flap still horizontal, until the forward end of the carrier is near its seat in the forward end of the receiver.

(e) Slide the carrier forward into its seat and pressing breechblock flat on the tang, slide the breechblock forward gently until it moves down to the locked position, allowing the flap of the carrier to rotate downward towards top of receiver. Do not force.

(f) Secure the carrier by means of the carrier stop, inserted vertically with pin forward and down into its T-slot in the inner wall of the receiver. With a screwdriver, push forward on the carrier cam to allow the pin on the stop to seat properly in its recess in the carrier cam.

(2) TRIGGER GUARD GROUP.

(a) With the receiver still upside down, slide the guard onto the rear of the receiver so that it mates with the grooves in lower rear face of the receiver and with the tang, and the tang and guard screw holes in guard and receiver are in alinement.

(b) Turn receiver and replace tang screw and turn in part way. Then replace guard screw from right side, catch threads, and turn until check screw cut is in position; then replace check screw. Then turn tang screw in tightly, aline check screw cut, and replace check screw.

(3) BUTT STOCK.

(a) Slide the butt stock onto the trigger guard tang and receiver until it mates evenly and flush.

(b) Insert stock bolt and washer into bolt aperture in butt stock, catch threads and screw down tightly.

(c) Replace butt plate and screw on snugly. Do not force screws, as composition plate may crack or wood screw threads strip out.

TM 9-285
41

REMINGTON SHOTGUN, 12-GAGE, M10

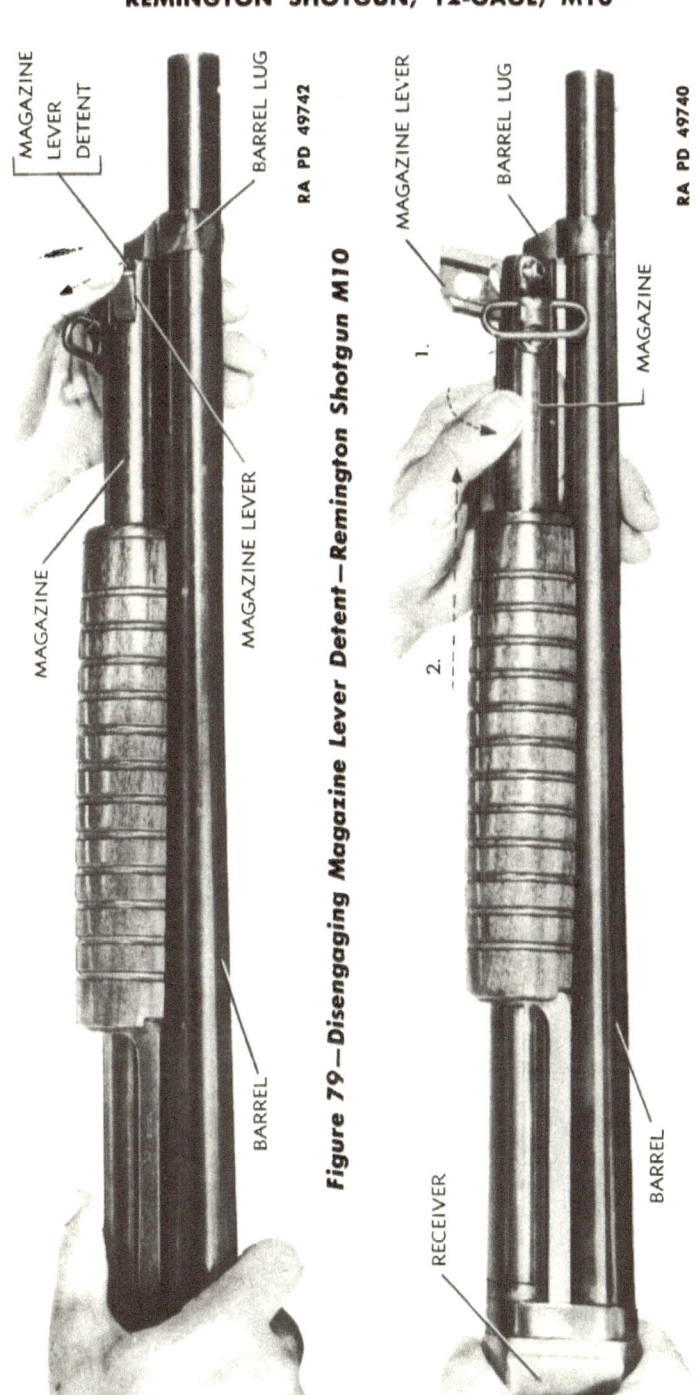

Figure 79—Disengaging Magazine Lever Detent—Remington Shotgun M10

Figure 80—Disengaging Magazine from Receiver—Magazine Lever in Disengaged Position—Remington Shotgun M10

TM 9-285

SHOTGUNS, ALL TYPES

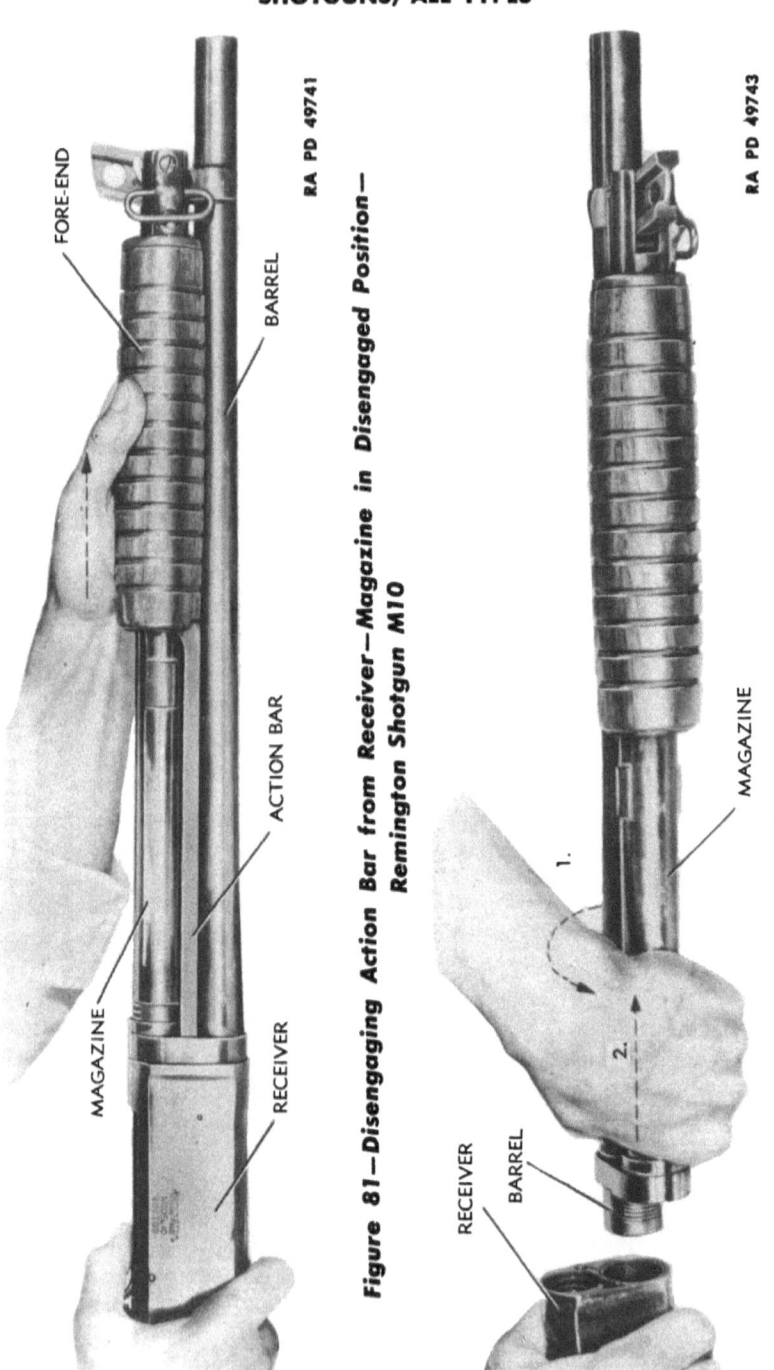

Figure 81—Disengaging Action Bar from Receiver—Magazine in Disengaged Position—Remington Shotgun M10

Figure 82—Barrel Disengaged from Receiver—Remington Shotgun M10

REMINGTON SHOTGUN, 12-GAGE, M10

(4) BARREL, MAGAZINE AND ACTION BAR GROUP.

(a) Move breechblock forward to locked position.

(b) Inspect bore of barrel for foreign matter, and then slide action bar fully forward to clear the rear face of the barrel yoke.

(c) Holding group with magazine facing to left, insert the threaded end of the barrel into the barrel opening in the receiver as far as it will go. Then turn the group counterclockwise ¼ turn until the barrel yoke is stopped by the yoke stop screw projecting from the right side of the forward face of the receiver. (The flat spot on the left side of the extreme rear end of the barrel is for the purpose of camming up the extractor when the barrel is pushed into the receiver. This flat should be up, in line with extractor, when inserting the barrel. The cut in the top of extreme rear end of the barrel is for the purpose of furnishing clearance for the extractor when barrel is assembled).

(d) Slide the action bar into the receiver to engage with the breechblock. Depress action bar lock button and move action bar and breechblock all the way to the rear. Then with magazine lever vertically in line with barrel and receiver, slide the magazine into the receiver as far as it will go and, using lever, turn magazine ¼ turn counterclockwise. Then rotate lever into magazine plug until retained by detent.

(e) Test the assembly by reciprocating the action bar several times with the action bar lock button depressed. Then, test for locking of breechblock by removing pressure from the lock button and pushing fore end all the way forward to lock the breechblock.

42. FIELD INSPECTION.

a. With the gun completely assembled, test the mechanism for proper functioning. Fired shells may often be used for testing, where dummy shells are not available, by turning in the uncrimped end so that the length of the shell approximates that of the live shell. Use of live shells for testing is prohibited.

CAUTION: Be sure gun is fully unloaded before inspection.

b. **Operate the Gun as Follows:**

(1) With the breechblock locked and the cocking head cocked, depress the action bar lock button protruding through the right wall of the receiver. Pull fore end smartly and fully to the rear and then push smartly and fully forward. Reciprocate fore end thus several times to test smoothness of action.

(2) Retract fore end as in (1) above, then release action bar lock button and push fore end smartly and fully forward to lock the breechblock. Then attempt to retract the fore end. The fore end should not retract.

SHOTGUNS, ALL TYPES

(3) Pull the trigger thus allowing the cocking head to move forward to the fired position and attempt to retract the fore end. The fore end should retract.

(4) Retract the fore end fully and then push forward until the breechblock is fully forward but not raised to the locked position. Then pull the trigger to release the cocking head. The trigger should not release the cocking head until the breechblock is fully locked. (The trigger should not contact the sear until the breechblock is locked).

(5) Place two or more dummy or fired shells in the magazine and work through the action to test gun for feeding, loading, extraction and ejection of shells. The second shell should not leave the magazine until the first shell has been ejected.

NOTE: Fired shells will not work through the action as easily as live or dummy shells as they are somewhat deformed through being fired. Therefore, allowance should be made for friction and smoothness of action in positioning the shell.

(6) With the breechblock locked and the cocking head in the cocked position, push the safety slide all the way to the rear and attempt to pull the trigger. The trigger should not pull.

(7) Push the safety slide all the way forward and attempt to pull the trigger. The trigger should pull and the cocking head be released to fire the gun.

c. When gun does not operate and function smoothly and properly when tested as above, damaged or improperly assembled parts are indicated as follows:

(1) ACTION BAR STICKS. May be due to bent action bar, burred or broken cam lug on bar or foreign matter in breechblock camming aperture, dented magazine tube or damaged action bar lock.

(2) BREECHBLOCK DOES NOT LOCK. May be due to worn or broken action bar cam lug, broken or missing breechblock supporting pins in receiver, foreign matter on face of breechblock, or in extractor cut or in front of locking shoulder in receiver.

(3) BREECHBLOCK DOES NOT UNLOCK. May be due to worn or broken action bar lug, broken or missing breechblock supporting pins, damaged action bar lock or breechblock latch, or foreign matter in mechanism.

(4) COCKING HEAD DOES NOT COCK PROPERLY OR SLIPS. May be due to worn or broken cam lug on action bar, worn sear nose, broken sear or sear spring, bent trigger, broken main spring, or foreign matter in mechanism.

(5) COCKING HEAD DOES NOT RELEASE WHEN TRIGGER IS PULLED.

REMINGTON SHOTGUN, 12-GAGE, M10

May be due to bent trigger, broken firing pin or main spring, damaged sear, or foreign matter in mechanism.

(6) ACTION BAR DOES NOT FUNCTION. May be due to broken action bar or cam lug, broken action bar lock or cocking head.

(7) ACTION BAR LOCK DOES NOT FUNCTION. May be due to broken action bar lock, broken or missing lock shoe spring, broken button spring or spring retainer, or foreign matter in mechanism.

(8) CARRIER DOES NOT FUNCTION. May be due to broken carrier, missing carrier friction spring, burred or worn cam on rear of carrier, damaged carrier stop, or foreign matter in mechanism.

(9) SHELL IS NOT EXTRACTED OR EJECTED. May be due to worn or broken extractor, broken extractor spring, or bent or broken ejector spring. If ejector spring is bent or too weak, it will not function properly.

(10) TWO SHELLS FED INTO RECEIVER AT ONCE. May be due to bent or broken carrier shell retainer spring, worn shell retainer collar, or worn cam on rear of carrier.

(11) SHELL STICKS IN MAGAZINE. May be due to corroded or bent follower, kinked or broken magazine spring, or dents, foreign matter or rust in magazine tube.

(12) SHELLS DO NOT CHAMBER PROPERLY. May be due to damaged carrier, as above, or stiff or bent ejector spring. The latter defect is usually the cause.

(13) SAFETY SLIDE DOES NOT FUNCTION. May be due to burs on slide, broken slide spring, or foreign matter in mechanism.

d. Inspect barrel and test trigger pull (par. 3 n).

e. In addition to inspection of the gun for operation and functioning, the gun should be inspected generally for condition, and defects noted. Attention should be directed to such defects as cracked wooden parts, cracked or deformed metal parts, dented magazine tube, loose screws and pins, loose or binding parts or assemblies, loose barrel or magazine, loose stock or butt plate, rust, dents, burs, or excessive wear of parts. If defects are such that early malfunction of the gun is indicated, the gun should be turned over to ordnance personnel for inspection and correction.

f. Where defects and malfunctions cannot be remedied by cleaning, lubrication, and simple adjustments of assembly, which lie within the scope of using troops, the gun should be turned over to ordnance personnel for a thorough inspection, correction and/or repair.

g. Removal of burs on working parts, trigger adjustments, and like corrections should not be attempted by using troops as stoning of parts

SHOTGUNS, ALL TYPES

must be exacting, the angle of the faces concerned must not be changed, and volume of metal must not be materially reduced.

h. A loose barrel, which shakes when assembled, may be due to improper assembly caused by not inserting barrel far enough into receiver before mating interrupted threads of barrel and receiver, or it may be due to worn parts necessitating adjustment or replacement. If due to latter cause, gun should be turned over to ordnance personnel for correction or repair.

i. If shell appears unnecessarily loose in chamber with breechblock locked, the guns should be turned over to ordnance personnel to be checked for headspace.

j. Adjustment and maintenance of the gun in the case of using troops is limited to the removal and replacement of the parts and groups of parts (pars. 40 and 41), together with cleaning, lubrication, and such adjustments as are necessary in assembling the gun as outlined.

43. CLEANING AND LUBRICATION.

a. Cleaning, oiling, and lubrication of this gun may be accomplished in a manner similar to that prescribed for the Winchester Gun M97 (par. 11). Attention should be given to corresponding parts and surfaces when lubricating. Barrel magazine and fore end group should be removed from receiver of take-down gun when cleaning the bore. With bayonet attachment assembled to the gun, the bore should be cleaned from the muzzle end, and a rag stuffed into the receiver to protect the action while cleaning (par. 3 i).

b. Points to lubricate are:
(1) Action bar opening in forward end of receiver.
(2) Action bar cam lug.
(3) Action bar lock pin, nose of lock, and shoe hinge.
(4) Carrier bearing points.
(5) Cocking head.
(6) Breechblock latch.
(7) Trigger pin.
(8) Safety slide.
(9) Carrier cam aperture in breechblock.
(10) Outer surface of magazine tube where action bar tube bears.

c. In very cold climates, oiling and lubrication should be reduced to a minimum. Only surfaces showing signs of wear should be lightly oiled. Refer to "Special Maintenance," section XI.

Section VII

REMINGTON SHOTGUN, 12-GAGE, M31

	Paragraph
Description	44
Data	45
Operation	46
Functioning	47
Removal of groups	48
Replacement of groups	49
Field inspection	50
Cleaning and lubrication	51

44. DESCRIPTION.

a. Identification marks on this gun are generally to be found as follows:

(1) Name of maker, gage, and chamber length are stamped on the top of the barrel near the rear end.

(2) Serial number of gun is stamped on the left side of the barrel near the rear end.

(3) Name of maker, model, and serial number of the gun are stamped on the left side of the receiver.

b. This gun (figs. 83 and 84) is a manually operated, repeating shotgun of the slide-action, hammerless and take-down type. The gun is so constructed that the barrel can easily be removed by disengaging the barrel lock, positioned in the magazine plug, from the barrel lug and unlocking the barrel from the receiver by disengaging interrupted threads on the rear end of the barrel from like threads in the receiver. This construction facilitates cleaning and transportation.

c. This gun is furnished in various grades having barrels of different lengths and degrees of boring and other modifications of design. Basically, however, the mechanism of all guns of this make and model are the same, except as indicated herein. For convenience the guns will be classified as three types: riot, skeet and sporting, although variations of these types may occur.

NOTE: The Remington Shotgun, Model 31, is manufactured in three grades: the 31-A (fig. 83) (standard), 31 skeet (fig. 85) and 31-TC (trap). The riot gun, described herein, is the 31-A (standard) grade; and the skeet gun, the 31 skeet grade. In this technical manual the guns are divided into three types only, as explained, which apply to length of barrel and boring irrespective of grade (par. 31).

SHOTGUNS, ALL TYPES

Figure 83—Right Side View—Riot Type—Remington Shotgun M31

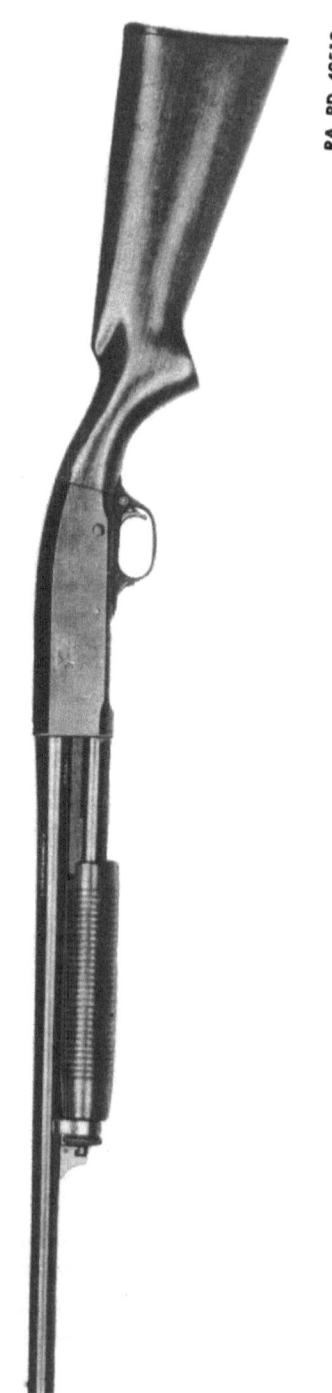

Figure 84—Left Side View—Riot Type—Remington Shotgun M31

TM 9-285
44

REMINGTON SHOTGUN, 12-GAGE, M31

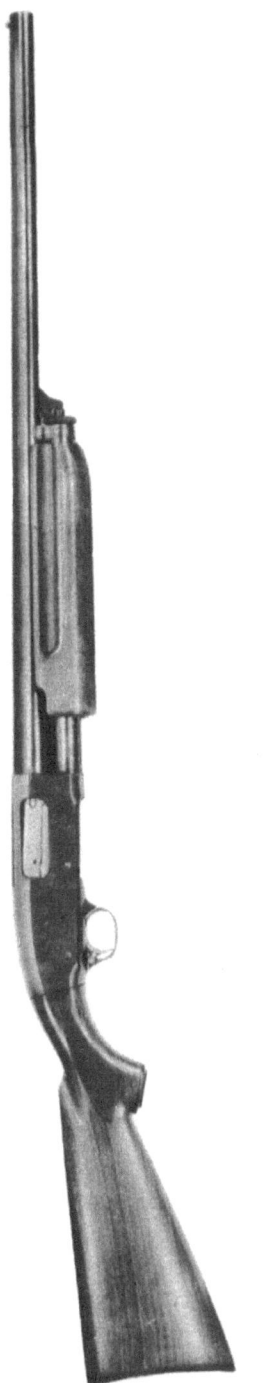

Figure 85—Right Side View—Sporting Skeet Type—Remington Shotgun M31

SHOTGUNS, ALL TYPES

(1) The riot-type gun (figs. 83 and 84) is usually furnished with a 20-inch plain barrel, bored full cylinder and may not be eqiupped with a bayonet attachment.

(2) The sporting skeet-type gun (fig. 85) is usually furnished with a 26-inch plain or ribbed barrel, bored improved cylinder, and is without bayonet attachment or sling.

(3) The sporting-type gun is similar to the skeet-type gun but is furnished with a 30-inch barrel, bored full choke, and is without bayonet attachment or sling.

d. **General Description** (figs. 86, 93 and 94).

(1) The stock of the gun is bolted to the rear end of the receiver, and the barrel locked to the forward end of the receiver by means of interrupted threads on barrel and in receiver, and fastened to the forward end of the magazine by means of a lug on the barrel and a stud on the magazine lock which is assembled to the magazine plug. The magazine tube is screwed into the receiver at manufacture and locked in position. The fore end and action bar group is mounted and operated on the magazine tube. The rear end of the action bar passes through the forward end of the receiver and engages with and operates the slide which in turn operates the breechblock and carrier, and cams back and cocks the hammer. The wooden grip of the fore end is thicker on the 31 skeet grade of gun (fig. 85) than in the 31-A grade gun (fig. 83) and extends some distance to the rear of the action bar tube when assembled (par. 50 j). The magazine lock of the skeet grade gun is held in engagement with the barrel lug by spring force while that of the 31-A grade gun is screwed into engagement.

(2) The receiver contains the operating mechanism and to its lower rear end is attached the trigger plate to which is mounted the firing mechanism and the action bar lock. The receiver is open at the bottom to permit loading and at the right side for ejection of the fired shell cases. The carrier pivots in the inside of the receiver and acts to raise the shell to bore line in loading. The right and left shell stops pivot in the forward end of the receiver to the rear of the magazine opening. The ejector is seated in a T-slot in the left wall of the receiver.

(3) The breechblock contains the right and left extractors, the firing pin, and their components and mates with and is operated by the slide by means of a hooked lug on the rear lower face of the breechblock seating in a like aperture in the top of the slide. When cammed up by the slide on the forward movement the top rear end of the breechblock seats in front of a locking shoulder cut in the top of the receiver to lock the breechblock in position. A rib staked in the top of the receiver acts

TM 9-285

REMINGTON SHOTGUN, 12-GAGE, M31

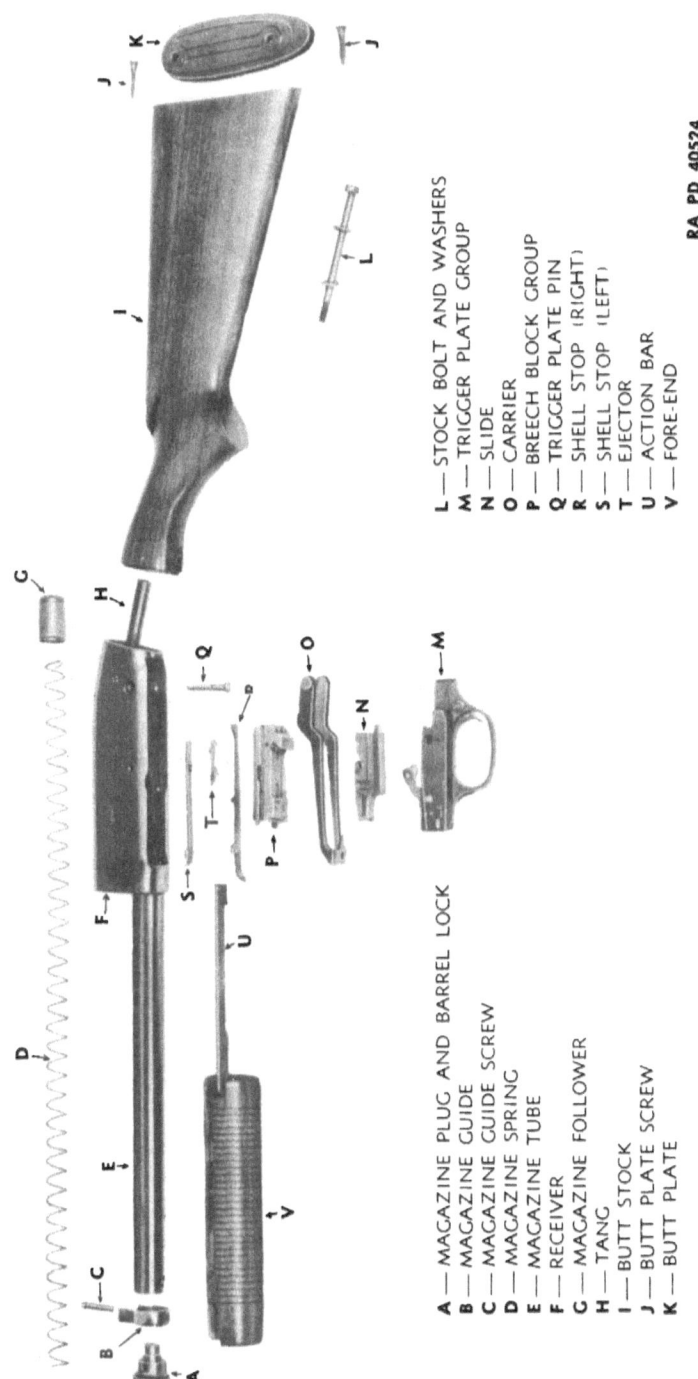

Figure 86—Disassembled View—(Without Barrel)—Riot Type—Remington Shotgun M31

A—MAGAZINE PLUG AND BARREL LOCK
B—MAGAZINE GUIDE
C—MAGAZINE GUIDE SCREW
D—MAGAZINE SPRING
E—MAGAZINE TUBE
F—RECEIVER
G—MAGAZINE FOLLOWER
H—TANG
I—BUTT STOCK
J—BUTT PLATE SCREW
K—BUTT PLATE
L—STOCK BOLT AND WASHERS
M—TRIGGER PLATE GROUP
N—SLIDE
O—CARRIER
P—BREECH BLOCK GROUP
Q—TRIGGER PLATE PIN
R—SHELL STOP (RIGHT)
S—SHELL STOP (LEFT)
T—EJECTOR
U—ACTION BAR
V—FORE-END

SHOTGUNS, ALL TYPES

as a guide for the breechblock to hold it in central alinement in the receiver while opposite the ejection opening.

(4) The slide is engaged with the rear end of the action bar by means of a lug on the bar seating in a mating notch in the left side of the slide. The slide operates in guideways cut in the side walls of the receiver.

(5) The trigger plate (figs. 91 and 92) contains the trigger and integral sear, hammer, mainspring, action bar lock and the safety, together with their components. In the left side of the trigger plate is the left shell stop spring plunger which furnishes spring action to the stop.

NOTE: The trigger plate group of the 31-A grade gun (figs. 83 and 84) and the 31 skeet grade gun (fig. 85) are different in design. The main difference is in the action bar lock. In the 31-A grade gun the lock is operated by a torsion spring seated in the lock and bearing on a lug on the hammer, while in the 31 skeet grade gun, it is operated by the mainspring follower which keys with the action bar lock spring housing seated in the lock. (The riot-type gun (figs. 83 and 84), is a 31-A grade gun, and the skeet-type gun (fig. 85) is a skeet grade gun.)

(6) The magazine, which is of the tubular type, is screwed into the receiver below the barrel and has a capacity of four shells loaded end to end. The shells are pressed together and fed into the receiver by the force of the magazine spring acting upon the magazine follower. Some grades of this gun are furnished in a three-shot model. In the three-shot model the magazine holds only two shells.

(7) The left-hand and right-hand shell stops are pivoted in the lower inner walls of the receiver. The left-hand stop is operated by the action bar, while the right-hand stop is operated by the slide to block and release the shells in the magazine.

(8) The action bar lock is pivoted in the trigger plate (figs. 91 and 92) and engages with the slide to block its rearward movement after it has locked the breechblock in position, thereby preventing premature unlocking of the breechblock. The lock is disengaged from the slide either by the action of the mainspring follower which moves forward with the hammer when the gun is fired (skeet grade), or by manual pressure on the lower end of the lock which projects through the floor of the trigger plate at the right forward end of the trigger plate guard bow. (In the 31-A grade gun the lock is disengaged by a lug on the hammer bearing upon the lock spring, or by manually pressing on the lock). In the rear of the action bar lock, the trigger lock is pivoted and operated by a plunger and spring, assembled in the action bar lock. The trigger lock blocks the trigger from being pulled until the action bar lock is engaged, thus preventing premature firing of the gun.

REMINGTON SHOTGUN, 12-GAGE, M31

(9) The trigger safety is positioned in the rear of the trigger plate guard bow and can be moved laterally to block and free the trigger, thus preventing or allowing its pulling to fire the gun.

45. DATA.

Gage of bore	12
Boring of barrel—riot type	Cylinder
Boring of barrel—sporting skeet type	Improved cylinder
Boring of barrel—sporting trap type	Full choke
Type of action	Slide
Type of firing mechanism	Hammerless
Type of magazine	Tubular
Capacity of magazine (5-shot model)	4 shells
Capacity of magazine (3 shot model)	2 shells
Length of barrel—riot type	20 in.
Length of barrel—sporting skeet type	26 in.
Length of barrel—sporting trap type	30 in.
Length of stock, receiver and magazine assembled (approx.)	33 in.
Length of assembled gun—riot type (approx.)	40 in.
Length of assembled gun—sporting skeet type (approx.)	46 in.
Length of assembled gun—sporting trap type (approx.)	50 in.
Weight of assembled gun—riot type, A grade (approx.)	6 7/8 lb
Weight of assembled gun—sporting skeet type and grade (approx.)	7 5/8 lb
Weight of assembled gun—sporting trap type (approx.)	7 3/4 lb

46. OPERATION.

a. The gun is operated by moving the fore end fully and smartly backward and forward. This action unlocks the breechblock, extracts and ejects the fired shell casing, cocks the hammer, transfers a live shell from the magazine to the chamber of the barrel, and relocks the breechblock behind the shell.

CAUTION: During operation, the muzzle of the gun should always be pointing at a safe spot.

b. Before the fore end can be retracted the action bar lock must be disengaged from the slide (fig. 95). To facilitate this the fore end should be moved slightly forward to allow the lock to disengage. If the gun has been fired and the hammer consequently forward (down), the only movement necessary is to reciprocate the fore end as above, as the descending hammer and consequent forward movement of the mainspring follower (31 skeet grade) or hammer lug (31-A grade) has already disengaged the lock. If, however, the hammer is in the cocked position, it is necessary to press back on the lower end of the lock, which

TM 9-285
SHOTGUNS, ALL TYPES

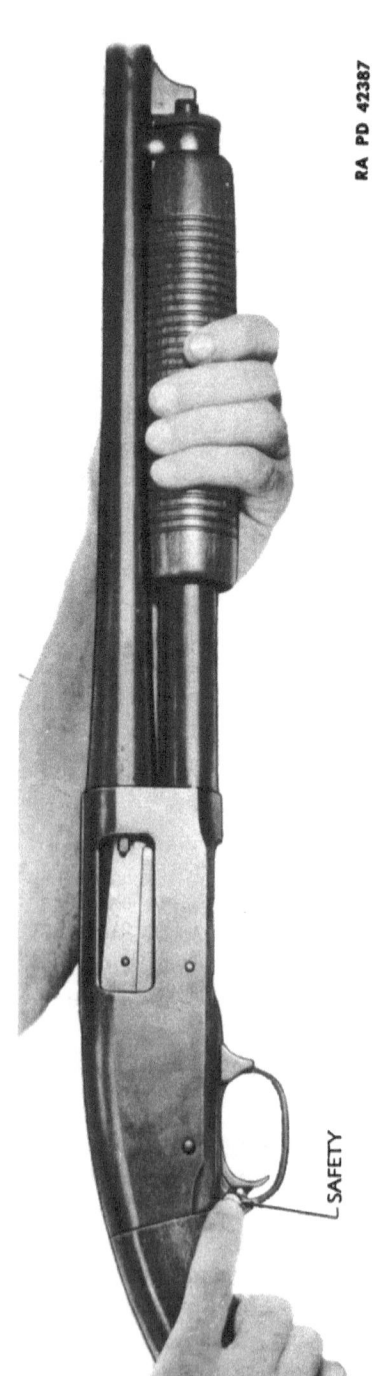

Figure 87 — Operating Safety — Remington Shotgun M31

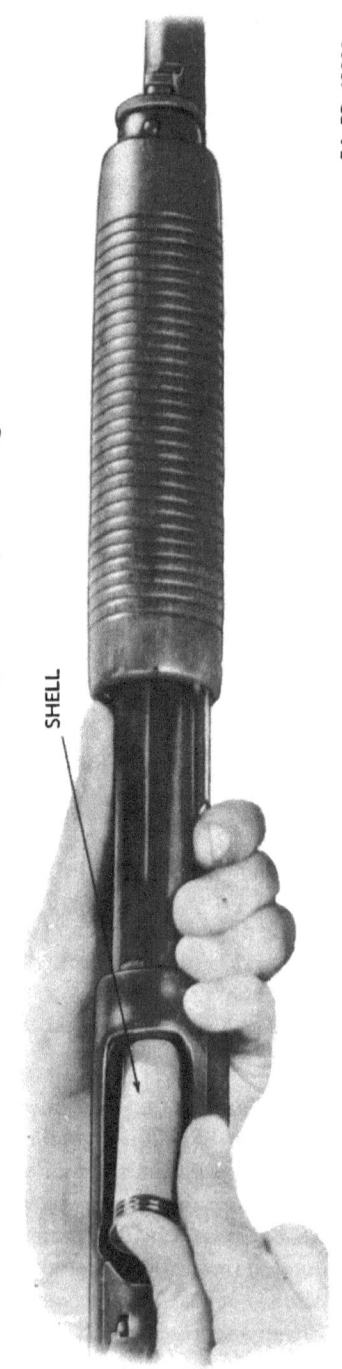

Figure 88 — Loading Magazine — Remington Shotgun M31

REMINGTON SHOTGUN, 12-GAGE, M31

is visible at the right side of the forward end of the trigger plate guard bow, before moving the slide forward and to the rear. When the gun is being fired as a repeater, the preliminary forward movement of the fore end is accomplished by the recoil of the gun away from the fore end, which is held by the operator.

CAUTION: During these operations, the finger should remain outside the trigger plate guard bow. Reciprocation of the fore end should be full and smart to insure extraction of the shell, cocking of the hammer, and complete locking of the breechblock and engagement of the action bar lock. Slamming of the mechanism, however, should be avoided.

c. With the gun loaded, locked, and cocked, the only movement necessary to fire the gun is the pulling of the trigger. Pressure on the trigger should be removed while the gun is being operated. The mechanism of this gun is such, however, that the hammer cannot be released until after the breechblock is locked and pressure on the trigger has been released and again applied. Thus premature firing is prevented.

d. The trigger safety, positioned in the rear of the trigger plate guard bow, operates laterally (fig. 87). When pushed to the left (red band showing), the trigger is free to be pulled and the gun fired. When pushed to the right, the trigger is blocked and cannot be pulled nor the gun fired. With a shell in the chamber, it is best to block the trigger by pushing the safety to the right unless the gun is to be fired immediately.

e. **To Load and Unload the Magazine.**

(1) To load the magazine (fig. 88), press a shell, nose first, into the rear end of the magazine against the magazine spring follower until it slips in front of and is retained by the left-hand shell stop. Load another shell in the same way by pressing it against the base of the first shell loaded, until four in all are loaded. Capacity of magazine is four shells for standard five shot gun, and two shells for three shot model. Loading should be done with the breechblock locked.

(2) To unload the magazine, press in on the left shell stop (with breechblock locked) and allow the shells to slip out one by one.

f. **To Load and Unload the Gun.**

(1) To load a shell from the magazine into the chamber, slide the safety to the safe position. Then disengage the action bar lock and reciprocate the fore end (par. 46 b). Another shell may then be loaded into the magazine. Test locking of the breechblock by attempting to retract the fore end. The fore end should not retract. Allow safety to remain in safe position unless gun is to be fired immediately.

(2) To unload the gun, slide the safety to the safe position and unload the magazine (par. 46 e (2)). Then, disengage the action bar lock and

SHOTGUNS, ALL TYPES

retract the fore end to extract and eject the shell in the chamber. Inspect magazine and chamber to make sure gun is fully unloaded.

g. **To Load the Chamber Only.** With the magazine empty, slide the safety to safe position, disengage the action bar lock, and retract the fore end. Then, insert a shell through the ejection opening in the receiver onto the carrier and push the fore end fully and smartly foreward to lock the breechblock behind the shell. Test locking of breechblock by attempting to retract the fore end. The fore end should not retract. Allow safety to remain in safe position unless gun is to be fired immediately.

47. FUNCTIONING.

a. As already explained, the functioning of the operating mechanism is accomplished by the reciprocation of the fore end. The rear end of the action bar is engaged with the slide by means of a lug on the bar mating with a notch in the slide. As the action bar is moved to the rear, it pushes the slide with it. The slide moves for a short distance independent of the breechblock, camming the rear end of the breechblock down and thus allowing the hooked cam lug on the lower rear end of the breechblock to drop down into a similarly shaped aperture in the top of the slide (fig. 89). This action unlocks the breechblock from the top of the receiver. From this point on, the slide pulls the breechblock with it to the rear. The breechblock extracts the fired shell from the chamber by means of the extractors in its forward end and the shell is knocked out of the receiver by the ejector positioned in the left side of the receiver wall. As the slide moves forward again, it pulls the breechblock with it and at the end of the forward movement cams up the rear end of the breechblock in front of the locking shoulder in the top of the receiver to lock the breechblock to the receiver (fig. 90).

b. During the rearward movement, the slide cams back the hammer, mounted in the trigger plate, and cams down the carrier. At the same time a shell is released from the magazine by cam action of a shoulder on the action bar with the left-hand shell stop. The shell is driven into the receiver by the force of the magazine spring and retained there by the carrier.

c. When the left-hand shell stop is cammed back by the action bar to release a shell from the magazine, the right-hand shell stop is cammed up by the slide to engage the following shell. This movement is reversed as the slide nears the end of its forward movement and the action bar allows the left-hand shell stop to resume engagement with the following shell which the right-hand shell stop has been blocking. Thus double feeding is prevented.

TM 9-285

REMINGTON SHOTGUN, 12-GAGE, M31

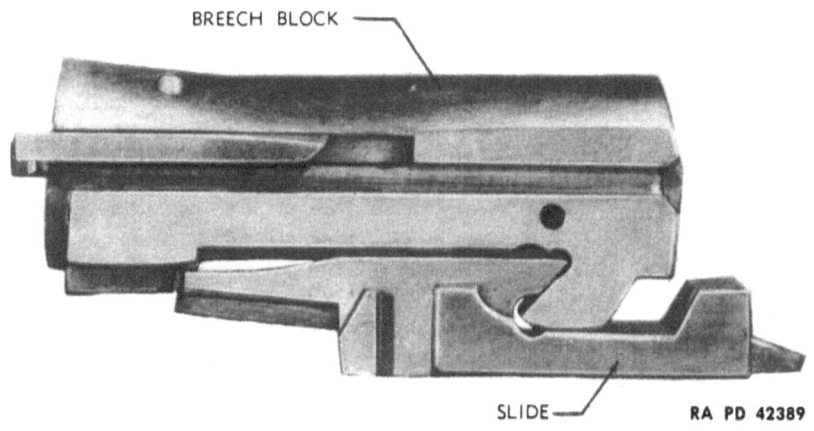

Figure 89 — Breechblock and Slide — Unlocked Position — Left Side View — Remington Shotgun M31

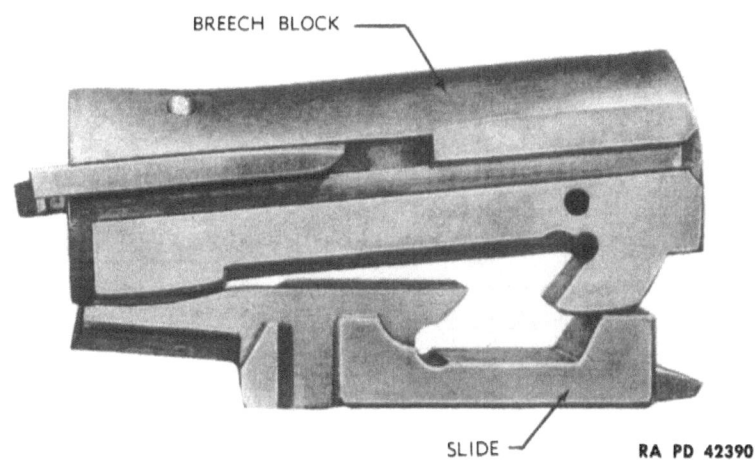

Figure 90 — Breechblock and Slide — Locked Position — Left Side View — Remington Shotgun M31

SHOTGUNS, ALL TYPES

d. As the slide is pulled forward by the action bar, it pulls the breechblock with it, as already explained. The slide cams up the carrier which lifts the shell in line with the chamber and the breechblock pushes the shell into the chamber.

e. As the hammer reaches the rearward position, it is caught and held by the nose of the trigger which acts as a sear. At the same time the trigger lock functioned by the slide lock engages with and blocks the trigger preventing its pulling. The trigger remains thus blocked until the slide has reached the extreme forward position and locked the breechblock. At this point the action bar lock, over which the slide has been riding, slips from under the slide and springs up in back of it to block

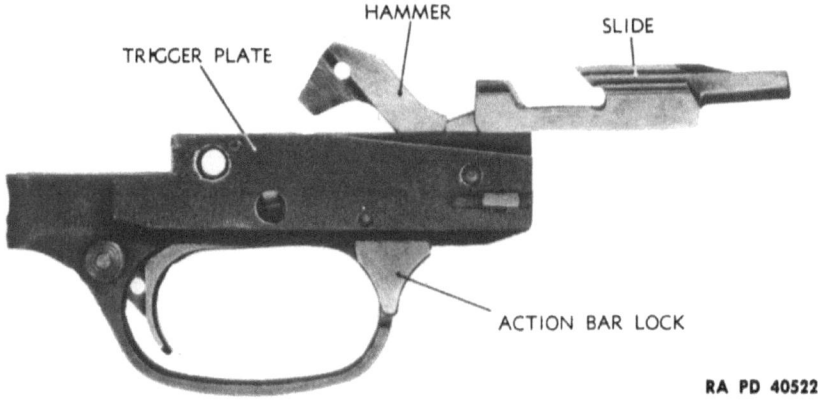

Figure 91—Slide and Trigger Plate Groups—Slide Riding Over Lock—Remington Shotgun M31

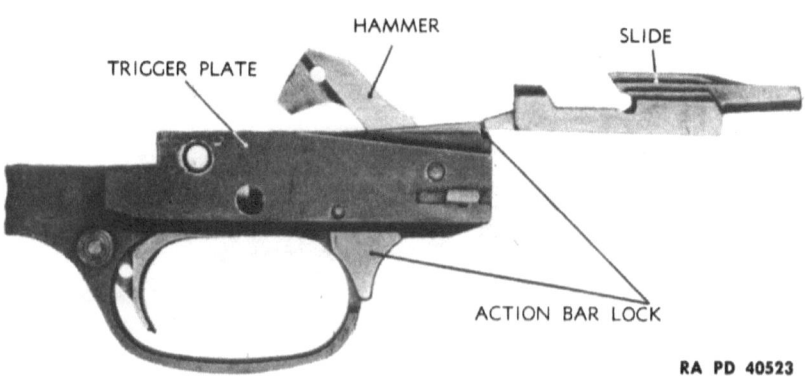

Figure 92—Slide and Trigger Plate Groups—Slide Blocked— Remington Shotgun M31

REMINGTON SHOTGUN, 12-GAGE, M31

rearward movement of the slide (figs. 91 and 92). As the lock springs up, it disengages the trigger lock from the trigger, which is now free to be pulled.

f. The hammer is released to fire the gun by pulling the trigger, which levers the nose of the trigger, which acts as a sear, down from engagement with the sear notch in the hammer.

48. REMOVAL OF GROUPS (figs. 86, 93 and 94).

a. Groups and parts should be removed and replaced in the order given below. Groups and parts when removed should be placed on a clean, flat surface and care observed to prevent loss of screws and small parts. Remove as follows:

(1) BARREL.

(a) Press action bar lock located just forward of trigger plate bow, to release slide by disengaging lock, and pull fore end all the way to the rear (fig. 95). This disengages extractors from rear end of barrel.

(b) Rest gun on butt plate with magazine forward. Grasp barrel lock, located at forward end of magazine, pull down (rearward) until free of stud and pin on rear face of barrel lug, and hold against spring pressure (fig. 96). In 31-A grade guns the barrel lock is threaded and is disengaged by turning counterclockwise.

(c) Turn barrel clockwise ¼ turn and pull from receiver (figs. 97 and 98). The magazine is screwed into the receiver at manufacture and should not be removed.

(d) Close action by pushing fore end all the way forward.

(2) TRIGGER PLATE GROUP.

(a) Push out trigger plate pin, located just above trigger, from right to left. Turn gun bottom up and, with breechblock in forward position, depress action bar lock, located at forward end of the trigger plate bow, to disengage lock from slide.

(b) With lock disengaged, slide trigger plate forward until right and left side lugs aline with mating slots in walls of receiver and pull group upward out of receiver. Then remove right and left shell stops from inner walls of receiver by pressing inward on pivot studs projecting through receiver walls near lower edge. Stops will usually shake out.

(3) SLIDE, BREECHBLOCK, AND CARRIER GROUP.

(a) With receiver bottom side up and magazine facing to left, move action bar rearward by pulling fore end to rear until rear face of small lug on rear end of slide is alined with forward edge of trigger plate pin hole in receiver.

TM 9-285
48

SHOTGUNS, ALL TYPES

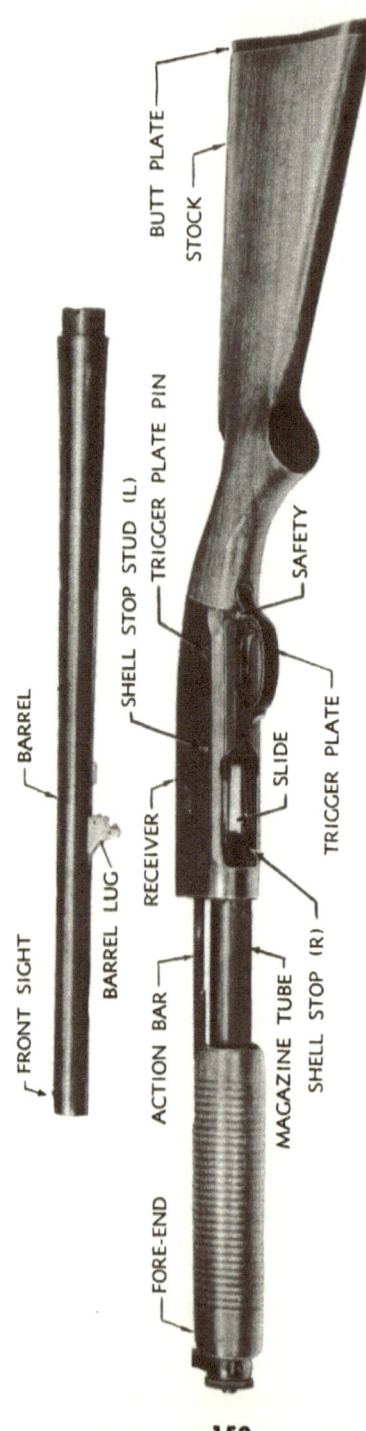

Figure 93—Gun Taken Down (Riot Type)—Left Side View—Showing Location of Parts—Remington Shotgun M31

TM 9-285

REMINGTON SHOTGUN, 12-GAGE, M31

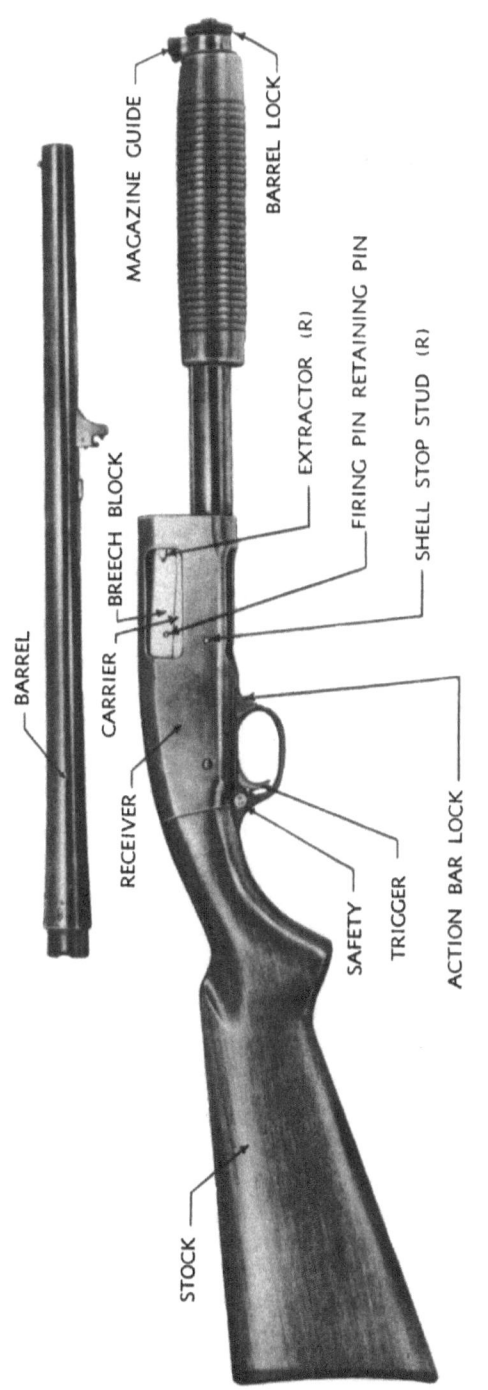

Figure 94—Gun Taken Down (Riot Type)—Right Side View—Showing Location of Parts—Remington Shotgun M31

SHOTGUNS, ALL TYPES

(b) Lift slide horizontally to clear action bar lug and move disengaged action bar fully forward.

(c) Move slide slightly forward to disengage hook lug from breechblock and holding slide up disengaged from breechblock, move breechblock fully forward to locked position, holding carrier down against breechblock to keep ejector depressed.

(d) With carrier resting on breechblock, move slide forward until forward end of guides on side faces of slide are ⅛ inch to rear of the trigger plate lug slots in right and left sides of receiver.

(e) Lift up right side of slide and swing slide up and out sidewise from receiver. This can be done without effort when slide is properly positioned and can be removed only when properly positioned. Do not force.

NOTE: To remove the slide, the forward end of the carrier must be resting upon the breechblock. In this position, it clears and presses the ejector into its seating groove. If the carrier slips up from the breechblock, it cannot be depressed again until the ejector is pressed out of the way of the carrier arm. This can be accomplished by inserting a small tool through the ejection opening in the side of the receiver and to the rear of the breechblock to press the ejector back into its seat, and thus allow the carrier to be depressed against the breechblock.

(f) Swing forward end of carrier upward out of receiver until carrier is perpendicular to receiver; then press rear ends together to disengage trunnions from receiver. Then lift carrier up out of receiver.

(g) Press ejector back into its seat, lift rear end of breechblock to disengage from locking shoulder in receiver, move to rear and lift from receiver. Then, remove ejector by pulling outward and forward. If ejector spring is loose in ejector, it should be removed to guard against loss.

49. REPLACEMENT OF GROUPS.

a. Groups and parts should be thoroughly cleaned, lightly oiled and lubricated, if necessary, before replacing. Replace as follows:

(1) SLIDE, BREECHBLOCK AND CARRIER GROUPS.

(a) With gun bottom side up and magazine to left, pull action bar fully forward.

(b) Insert the ejector into its T-shaped seat in the left inner wall of the receiver. If ejector spring has been disassembled, replace it in its seat in rear inner face of ejector so that expanded spring coil holds spring in seat. To seat ejector, insert T-end, with spring towards receiver wall, into undercut vertical groove in ejector seat and push ejector back and towards receiver until seated against spring force. Do not lever ejector in towards receiver until forward end is mating with longitudinal groove of seat. Ejector should seat easily. Do not force.

TM 9-285
49

REMINGTON SHOTGUN, 12-GAGE, M31

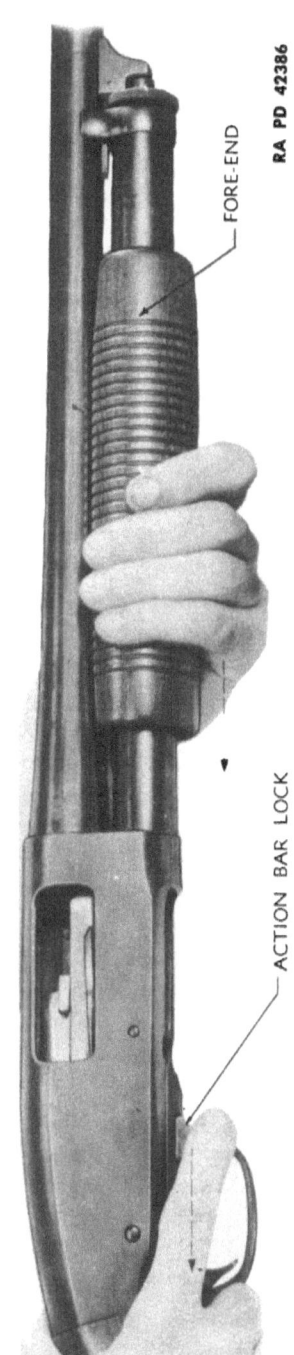

Figure 95—Disengaging Action Bar Lock and Retracting Fore End—Remington Shotgun M31

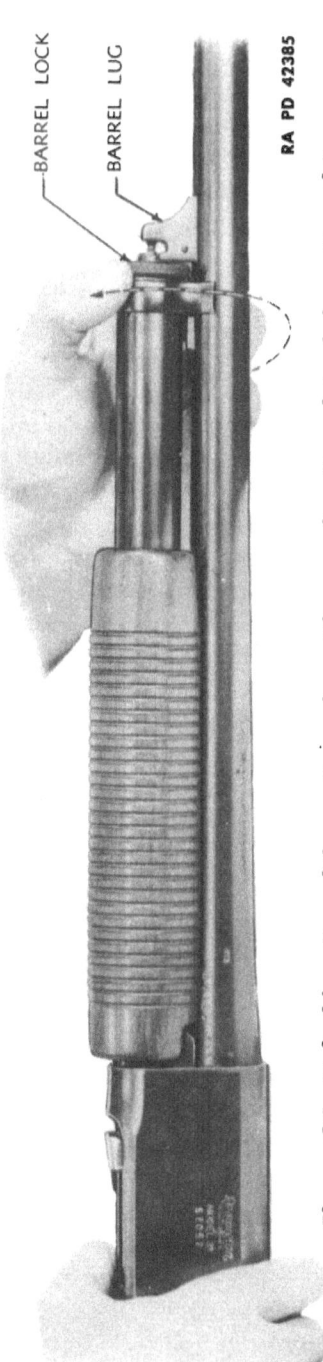

Figure 96—Unlocking Barrel from Magazine—Disengaging Barrel Lock from Barrel Lug—Remington Shotgun M31

TM 9-285
49

SHOTGUNS, ALL TYPES

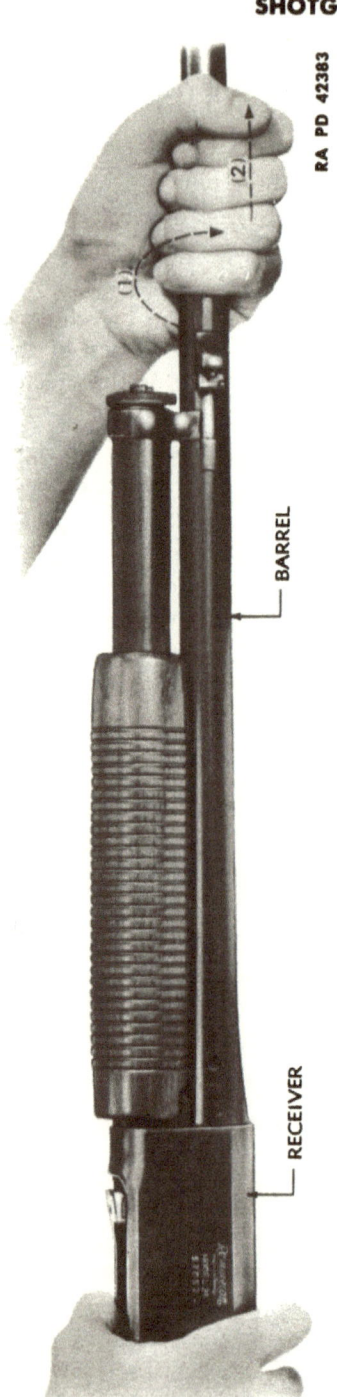

Figure 97 — Disengaging Unlocked Barrel from Receiver — Remington Shotgun M31

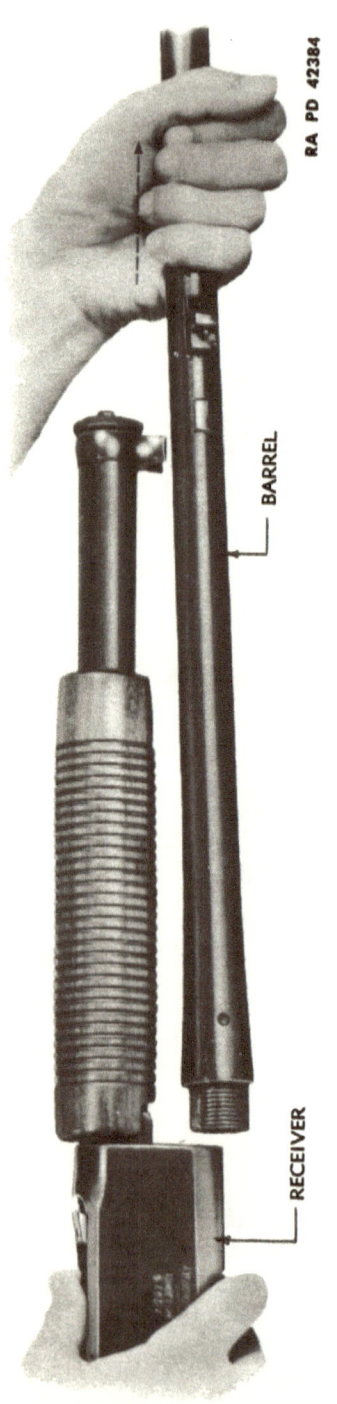

Figure 98 — Barrel Disengaged from Receiver — Remington Shotgun M31

REMINGTON SHOTGUN, 12-GAGE, M31

(c) Holding ejector in position, insert breechblock horizontally into rear of receiver so that rounded face is down toward receiver and extractors pointing to left. Push breechblock all the way forward to the locked position to hold ejector in place. Be sure slot in top of breechblock mates with rib in top of receiver to hold it centrally alined.

(d) Insert forked end of carrier vertically into rear of receiver, with flat of forward end facing to right (rear), and engage trunnions in their seats in rear, inner walls of receiver.

(e) Swing forward end of carrier downward until the breechblock lies between the side arms of the carrier and the forward end of the carrier is resting on the breechblock. To accomplish this the ejector must be depressed so that the carrier arm can clear it.

(f) With flat face of slide up and small lug on rear pointing to right insert slide at an angle, into receiver, left (far) guide first and down, at a point where the forward ends of the guides on either side of slide will be $1/8$ inch to rear of trigger plate lug mating slots in receiver walls.

(g) Mate the left (far) guide with its guideway in the receiver and then press the right (near) guide down into its guideway. This seating of the slide is easily accomplished if correctly done. Do not force. If seating is difficult, manipulate slide by raising left (far) guide slightly until it slips into its guideway; the right (near) guide will then easily slip down and into its guideway and the slide will lie flat.

NOTE: To replace the slide, the forward end of the carrier must be resting upon the breechblock. In this position, it clears and presses the ejector into its seat. If the carrier slips up from the breechblock, it cannot be depressed again until the ejector is pressed out of the way of the carrier arm. This can be accomplished by inserting a small drift through the ejector opening in the side of the receiver and to the rear of the breechblock to press the ejector into its seat, and thus allow the carrier to be pressed down against the breechblock.

(h) With the slide seated in its guideways, move it to the rear until rear face of small rear lug is in line with forward edge of trigger plate pin hole. Then raise the rear end of the breechblock to free it from the locking shoulder in the top of the receiver and move it to right (rear) under the slide until cam lug on breechblock slips into camming aperture in slide and slide rests flat on breechblock. Slide is easily lifted when positioned as above.

(i) Depress the forward end of carrier slightly to clear the end of the action bar and move action bar to rear (right) by means of the fore end until the lug on rear end of bar strikes the slide. Then lift slide off breechblock and move action bar to rear until lug on bar slips into notch in slide. Press slide down to mate with bar lug and breechblock and holding slide in contact with breechblock, move to forward position

SHOTGUNS, ALL TYPES

by moving fore end forward (left). Do not attempt to operate slide unless held down by finger until trigger plate is in position as it will jump up and jam in guideway.

(j) Seat right shell stop (longer of two) and left shell stop into their grooves in the lower inside faces of the receiver walls so that the curved ends point forward and the pivot studs seat in stud holes in receiver. Press stops flush with receiver. The trigger plate group can then be replaced (par. 49 a (2)).

(2) Trigger Plate Group.

(a) With breechblock in the locked position, receiver bottom side up, and shell stops in place as described in a (1) above, cock hammer and insert trigger plate horizontally into receiver so that lugs on forward end of trigger plate fit into lug slots in lower edges of receiver walls. Press plate down horizontally into receiver, exercising care to hold shell stops flush with receiver walls while seating plate. When plate is seated, slide to rear until pin holes in plate and receiver aline. If replaced correctly, the trigger plate seats easily. Do not force.

(b) With trigger plate seated flush in receiver, pin holes in receiver, and plate alined, insert trigger plate pin, small end first, into pin hole in left side of receiver. Push pin through until head is seated flush in counterbore. The trigger plate pin is spring retained.

(3) Barrel.

(a) Press action bar lock, positioned in forward end of trigger plate bow, to disengage lock from slide, and move slide part way to rear by pulling fore end to rear to prevent interference of extractors with barrel when inserting barrel in receiver. If fore end is pulled all the way to rear, it will interfere with barrel when seating in receiver.

(b) Rest gun upon butt plate in vertical position with magazine forward and grasp barrel lock. Pull spring-type barrel lock down, and hold against spring pressure. If barrel lock is of screw type, retract fully by turning counterclockwise.

(c) Inspect bore for foreign matter and insert threaded end of barrel into barrel aperture of receiver, with barrel lug pointing to left, until barrel shoulder is stopped by forward face of receiver, with interrupted threads in barrel in position to mate with those in receiver.

(d) Turn barrel counterclockwise until threads on barrel and in receiver engage and barrel lug stops against barrel lock.

(e) Release barrel lock to spring forward, and turn until stud and pin on barrel lug fully engage with their seats in the lock. If screw-type lock, turn clockwise until fully engaged.

(f) Close action by pushing fore end fully forward. Operate the gun to test assembly and locking of breechblock and slide.

REMINGTON SHOTGUN, 12-GAGE, M31

50. FIELD INSPECTION.

a. With the gun completely assembled, test the mechanism for proper functioning. Fired shells may often be used for testing where dummy shells are not available, by turning in the uncrimped end so that the length of the shell will approximate that of a live shell. Use of live shells for testing is prohibited.

CAUTION: Be sure gun is fully unloaded before inspection.

b. **Operate the Gun as Follows:**

(1) With the breechblock locked and the hammer cocked, press back on the rear end of the action bar lock, positioned in forward end of trigger plate guard bow. Push fore end forward slightly; then, pull smartly and fully to the rear and then push smartly and fully forward. Reciprocate fore end thus several times to test smoothness of action.

(2) Retract fore end as in (1) above; then release action bar lock and push fore end smartly forward to lock the breechblock. Then attempt to retract the fore end. The fore end should not retract.

(3) Pull the trigger, thus allowing the hammer to move forward to the fired position and attempt to retract the fore end. The fore end should retract.

(4) Retract the fore end fully and then push forward until the breechblock is fully forward but not raised to the locked position. Then pull the trigger to release the hammer. The trigger should not pull nor release the hammer until the breechblock is fully locked.

(5) Hold firmly back on the trigger while moving the fore end forward to lock the breechblock. The hammer should not be released to fire gun even after the breechblock is fully locked until pressure is released from the trigger and reapplied.

(6) Place two or more dummy or fired shells in the magazine and work through the action to test gun for feeding, loading, extraction and ejection of shells. The second shell should not leave the magazine until the first shell has been ejected from the receiver.

NOTE: Fired shells will not work through the action as easily as live or dummy shells as they are somewhat deformed through being fired. Therefore, allowance should be made for friction and smoothness of action in positioning the shell.

(7) With breechblock locked and hammer in cocked position, slide trigger safety all the way to the right and attempt to pull the trigger to release the hammer. The trigger should not pull nor the hammer be released.

(8) Slide safety all the way to the left (so that red band shows) and attempt to pull the trigger. The trigger should pull and the hammer be released to fire the gun.

SHOTGUNS, ALL TYPES

c. When gun does not operate and function smoothly and properly when tested as above, damaged or improperly assembled parts are indicated as follows:

(1) ACTION BAR STICKS. May be due to bent action bar, or burred lug on rear end of bar.

(2) BREECHBLOCK DOES NOT LOCK. May be due to foreign matter on face of breechblock, on locking shoulder in top of receiver, or in extractor grooves in receiver or rear end of barrel, or burs on locking lugs on breechblock or slide, or foreign matter in lug aperture in slide.

(3) HAMMER DOES NOT COCK PROPERLY OR SLIPS. May be due to wear or foreign matter in sear notch in hammer or on nose of sear, improperly assembled or broken trigger spring, improperly assembled action bar lock, broken or improperly assembled hammer link, or broken mainspring.

(4) FIRING PIN DOES NOT RETRACT IN BREECHBLOCK. May be due to broken firing pin spring or firing pin, or foreign matter in aperture.

(5) ACTION BAR LOCK DOES NOT FUNCTION. May be due to bent or burred action bar lock, broken spring, improperly assembled lock spring housing (31 skeet grade gun) or mainspring follower, displacement of hammer link, bent or broken lock spring (31-A grade gun), or burred or worn nose of lock or contacting lug on rear of slide.

NOTE: Action bar lock spring housing (31 skeet grade gun) operates in a guideway in right wall of trigger plate. The lock is tilted by the housing which also slides in a groove in the lock. The housing is functioned by the mainspring follower to which it is keyed (par. 44 d (8)).

(6) SHELL IS NOT EXTRACTED OR EJECTED. May be due to worn, broken or burred extractors, broken or missing extractor springs, broken ejector, or broken or missing ejector spring.

(7) TWO SHELLS FED INTO RECEIVER AT ONCE. May be due to bent or broken right shell stop, or foreign matter in shell stop seating grooves in receiver. If first shell is not retained in magazine, the left shell stop plunger spring (in trigger plate) may be broken.

(8) SHELL STICKS IN MAGAZINE. May be due to corroded or bent magazine spring follower, dented tube, broken or kinked spring or corrosion or foreign matter in tube.

(9) TRIGGER SAFETY STICKS. May be due to burred safety or trigger web, or broken or improperly assembled detent spring or ball.

(10) TRIGGER LOCK DOES NOT FUNCTION. May be due to broken or missing plunger or spring, or damaged lock.

d. Inspect barrel and test trigger pull (par. 3 n).

e. In addition to inspection of the gun for operation and functioning, the gun should be inspected generally for condition and defects noted. Attention should be directed to such defects as cracked wooden parts,

REMINGTON SHOTGUN, 12-GAGE, M31

cracked or deformed metal parts, dented magazine tube, loose screws and pins, loose or binding parts or assemblies, loose barrel or magazine, loose stock or butt plate, rust, dents, burs, or excessive wear of parts. If defects are such that early malfunction of the gun is indicated, the gun should be turned over to ordnance personnel for inspection and correction.

f. Where defects and malfunctions cannot be remedied by cleaning, lubrication, and simple adjustments of assembly, which lie within the scope of using troops, the gun should be turned over to ordnance personnel for a thorough inspection, correction, and/or repair.

g. Removal of burs on working parts, trigger adjustments, and like corrections should not be attempted by using troops as stoning of parts must be exacting, the angle of the faces concerned must not be changed, and volume of metal must not be materially reduced.

h. A loose barrel, which shakes when assembled, may be due to improper assembly caused by not inserting barrel far enough into receiver before mating interrupted threads of barrel and receiver, or it may be due to worn parts necessitating adjustment or replacement with new parts. If due to worn parts, the gun should be turned over to ordnance personnel for correction or repair.

i. If shell appears unnecessarily loose in chamber with breechblock locked, the gun should be turned over to ordnance personnel to be checked for headspace.

j. The wooden handle of the fore end of the 31 skeet grade gun extends for some distance to the rear of the action bar tube. This portion of the fore end sometimes becomes cracked through interference with the carrier when operated. If the inner, rear face of the wooden handle shows signs of having been struck by the carrier, the gun should be turned over to ordnance personnel for correction in accordance with instructions contained in **TM 1285**.

k. The rear end of the wooden handle of the 31-A grade gun, while considerably shorter than that of the 31 skeet grade, projects about ½ inch beyond the action bar tube and tapers to a thin edge which is apt to crack. If cracking is evident, the gun should be turned over to ordnance personnel for removal of the overhanging portion of the handle to prevent further cracking of the handle.

l. Adjustment and maintenance of the gun in the case of using troops is limited to the removal and replacement of the parts and groups of parts as outlined in paragraphs 48 and 49, together with cleaning, lubrication, and such adjustments as are necessary in assembling the gun as outlined.

SHOTGUNS, ALL TYPES

51. CLEANING AND LUBRICATION.

a. Cleaning, oiling, and lubrication may be accomplished in a manner similar to that described for the Winchester Gun M97 as outlined in paragraph 11. Attention should be given to corresponding parts and surfaces when lubricating.

b. The barrel should be removed from the receiver for cleaning the bore, and for thorough cleaning, the groups should be removed and the parts and assemblies cleaned, oiled, and lubricated as directed.

c. Points to be lubricated are:
(1) Action bar opening in forward end of receiver.
(2) Action bar guideway in receiver.
(3) Carrier trunnions.
(4) Slide guideways in receiver.
(5) Mainspring follower.
(6) Trigger and hammer pins.
(7) Slide lock spring housing guideway in right wall of trigger plate (skeet grade gun).
(8) Outer surface of magazine tube where action bar tube bears.

d. In very cold climates, oiling and lubricating should be reduced to a minimum. Only surfaces showing signs of wear should be lightly oiled. Refer to "Special Maintenance," section XI.

Section VIII

REMINGTON SHOTGUN, 12-GAGE, M11 AND SPORTSMAN

	Paragraph
Description	52
Data	53
Operation	54
Functioning	55
Removal of groups	56
Replacement of groups	57
Use of friction ring	58
Field inspection	59
Cleaning and lubrication	60

52. DESCRIPTION.

a. Identification marks on these guns are generally to be found as follows:

(1) Name of maker, gage size, and chamber length are stamped on the left side of the barrel near the rear end, and the serial number of the gun on the rear end of the barrel guide.

(2) The words "THE SPORTSMAN" are engraved on the right face of the breech bolt of this model gun.

(3) The name of maker and serial number of the gun are stamped on the left side of the receiver.

(4) Serial number of gun is stamped on the left face of the receiver tang and trigger plate tang.

b. The above guns are practically identical with the exception that the standard M11 Gun has a magazine capacity of four shells; the Sportsman Model, a capacity of two shells. The guns will, therefore, be treated as one gun and the Sportsman described, in this technical manual, with known differences noted. Any slight differences which may occur in the models due to slight variation of design or changes which may occur on account of dates of manufacture must be considered as such.

c. This gun (figs. 99 and 100) is an autoloading or semiautomatic shotgun, sometimes referred to in error as an automatic shotgun, of the take-down design. This gun is recoil-operated; the force, generated by the expanding powder gas acting against the breech bolt on the recoil movement together with the force generated by the compressed recoil and action springs on the counterrecoil movement, operates the mechanism of the gun to load, feed, extract and eject the shell and cock the hammer. The trigger must be pulled each time to fire the gun.

TM 9-285
52

SHOTGUNS, ALL TYPES

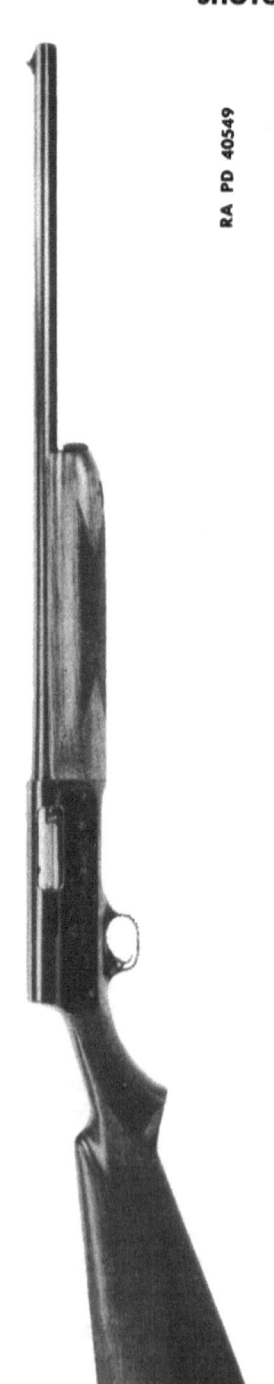

Figure 99—Right Side View—Sporting Skeet Type—Remington Shotgun, Sportsman

RA PD 40549

Figure 100—Left Side View—Sporting Skeet Type—Remington Shotgun, Sportsman

RA PD 40550

TM 9-285
52

REMINGTON SHOTGUN, 12-GAGE, M11 AND SPORTSMAN

d. The gun is so constructed that the barrel can easily be removed from the receiver by disengagement of the take-down screw (Sportsman) or by removal of the magazine cap (M11), and pulling fore end and barrel from the receiver. This construction facilitates cleaning and transportation.

e. This gun is furnished in various grades with barrels of different lengths and degrees of boring and may have other modifications of design. For convenience the guns (both models) will be classified as three types: riot, sporting skeet, and sporting trap, although variations of these types may occur.

(1) The riot-type gun is usually furnished with a 20-inch plain barrel, bored full cylinder. Autoloading type guns are not equipped with bayonet attachments.

(2) The sporting skeet-type gun (figs. 99 and 100) is usually furnished with a 26-inch plain or ribbed barrel, bored improved cylinder.

(3) The sporting trap-type gun is similar to the skeet-type gun but is furnished with a 30-inch barrel, bored full choke.

f. **General Description** (figs. 101, 108, and 109).

(1) The stock of the gun is fastened to the tang of the receiver by the tang screw passing through the tang of the trigger plate, through the stock grip, and threaded into the receiver tang. This screw is locked in place by a locking screw. To the rear end of the stock, a composition butt plate is fastened by two wood screws.

(2) The barrel has a semi-cylindrical extension threaded to the rear end, which slides into the forward end of the receiver and a ring guide brazed to the barrel tube, which slides over the magazine tube. The magazine tube is screwed into the receiver at manufacture and positioned by a stop screw. The recoil spring is mounted on the magazine tube together with a friction ring, friction piece, and spring. The positioning of these latter parts determine the amount of resistance (friction) set up against the force of recoil when the barrel is pulled to the rear by the breech bolt, when locked to it. The barrel is reciprocated within the receiver and upon the magazine tube by the breech bolt which is locked to it during the rearward movement, and again at the end of the forward movement by means of the locking block passing through an aperture in the barrel extension. A bead sight is driven into a ramp pinned and brazed to the muzzle end of the barrel and the ejector of the gun is riveted in a groove in the rear inner wall of the barrel extension.

(3) The magazine tube is screwed into the forward end of the receiver and contains the magazine spring and follower retained by a plug, screwed and staked into the forward end of the magazine (Sportsman), or spring retainer (M11).

TM 9-285

SHOTGUNS, ALL TYPES

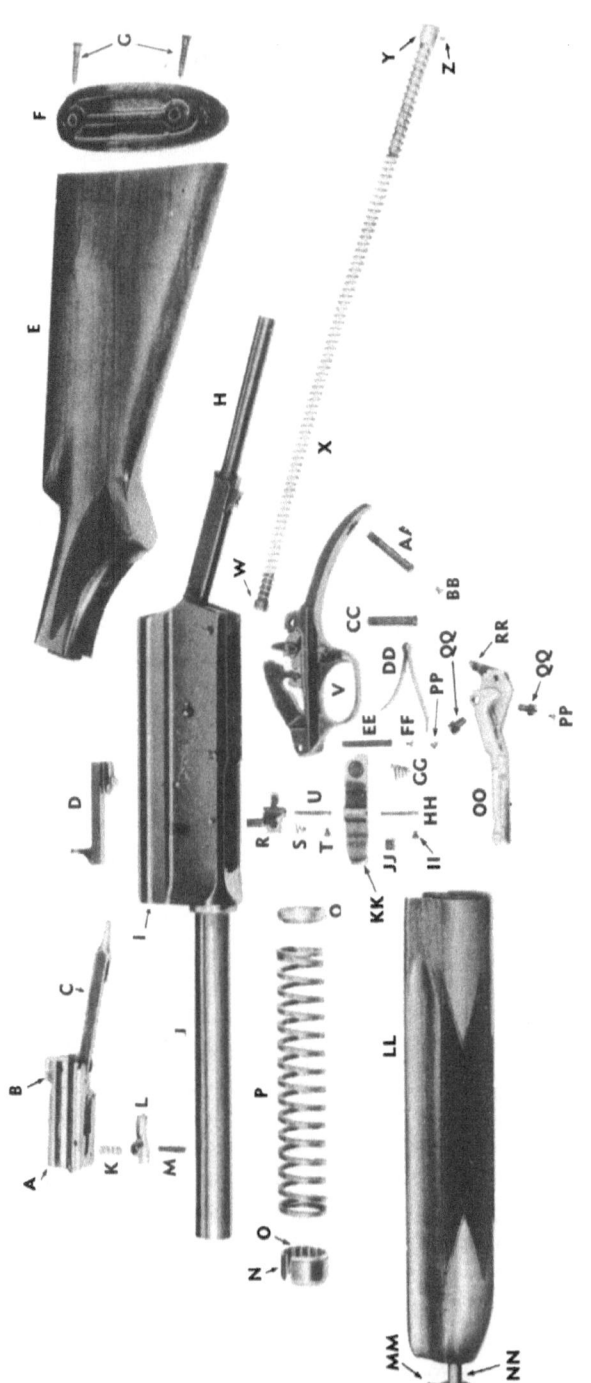

Figure 101 — Disassembled View (Without Barrel) — Remington Shotgun M11 — Sportsman

TM 9-285

REMINGTON SHOTGUN, 12-GAGE, M11 AND SPORTSMAN

A — BREECH BOLT
B — LOCKING BLOCK
C — LINK
D — OPERATING SLIDE
E — BUTT STOCK
F — BUTT PLATE
G — BUTT PLATE SCREW
H — ACTION SPRING TUBE
I — RECEIVER
J — MAGAZINE TUBE
K — LOCKING BLOCK LATCH SPRING
L — LOCKING BLOCK LATCH
M — LOCKING BLOCK LATCH PIN
N — FRICTION SPRING
O — FRICTION PIECE
P — RECOIL SPRING
Q — FRICTION RING
R — SHELL STOP
S — SHELL STOP SPRING
T — SHELL STOP PIN LOCK SCREW
U — SHELL STOP PIN
V — TRIGGER PLATE GROUP
W — ACTION SPRING FOLLOWER
X — ACTION SPRING
Y — ACTION SPRING PLUG
Z — ACTION SPRING PLUG PIN
AA — TANG SCREW
BB — TANG SCREW LOCKING SCREW
CC — TRIGGER PLATE SCREW
DD — CARRIER SPRING
EE — TRIGGER PLATE PIN
FF — TRIGGER PLATE PIN LOCKING SCREW
GG — CARRIER LATCH SPRING
HH — CARRIER LATCH PIN
II — CARRIER LATCH PIN LOCK SCREW
JJ — CARRIER LATCH BUTTON
KK — CARRIER LATCH
LL — FORE-END
MM — TAKEDOWN SCREW PIN
NN — TAKEDOWN SCREW
OO — CARRIER
PP — CARRIER SCREW LOCKING SCREW
QQ — CARRIER SCREW
RR — CARRIER DOG.

RA PD 40548A

Legend for Figure 101

SHOTGUNS, ALL TYPES

(4) The fore end slides over the magazine tube and is held in position by the take-down screw (Sportsman), or magazine cap (M11), and in turn limits the forward movement of the barrel. The magazine cap (M11) is locked in position by a spring plunger.

(5) The receiver contains the operating mechanism, and to its lower rear end is fastened the trigger plate to which is mounted the firing mechanism. The trigger plate is secured by a screw and a pin (with locking screw M11). The receiver is open at the bottom for loading and at the right side for ejection of the fired shell cases. To the inner rear wall of the receiver, a fibre cushion is riveted to cushion the breech bolt at the end of the rearward movement. The rear of the receiver is formed into a tang in which is pinned the action spring tube. The tang and tube extend into the stock when the gun is assembled. The action spring tube houses the action spring which is stopped from rearward movement by a wooden stop pinned in the rear end of the tube. The action spring follower is positioned in the forward end of the action spring, and against it bears the rear end of the link when assembled. The action spring forces the breech bolt forward on the counterrecoil stroke through the medium of the link which is pivoted to the locking block. The locking block slides radially in the bolt and is held in position by the firing pin which passes through bolt and locking block. It is this hinged construction of the link and locking block that enables the bolt to reciprocate horizontally while at an angle to the link and action spring. The vertical movement of the link operates the locking block latch positioned in the breech bolt, by which the locking block is blocked or freed to move radially in the bolt, when acted upon by the link.

(6) The carrier is pivoted in the rear of the receiver on two screws extending through the walls of the receiver and locked in place by locking screws. In the rear end of the carrier a dog is pivoted under spring tension supplied by a spring and plunger, positioned behind it in the carrier. The carrier is raised by means of the dog which bears on the lower face of the breech bolt, and lowered by action of the carrier spring, positioned in the receiver.

(7) The carrier latch is pivoted in a groove in the inner right side of the receiver, on a slotted pin secured by a stop screw, and is actuated by a coil spring, positioned between it and the receiver. A button bearing on the latch extends through the right wall of the receiver by which the latch can be manually disengaged from the carrier to free it.

NOTE: The pin and stop screw are replaced in some guns by a screw.

(8) The shell stop is similarly pivoted and actuated in the left side of the receiver and acts to check the fed shell from entirely entering the receiver until the proper time.

REMINGTON SHOTGUN, 12-GAGE, M11 AND SPORTSMAN

(9) The breech bolt (fig. 102) functions in guideways in the inner walls of the receiver, and houses the firing pin which operates longitudinally through the length of the bolt and the locking block, and is retracted into the bolt by the firing pin spring. The extractor, positioned in the forward right face of the bolt, is of the claw type and spring-actuated. The locking block functions radially in guideways in the breech bolt and is positioned by the firing pin which passes through it and the breech bolt. To the lower rear end of the locking block, the link is pivoted, by which the block is raised to lock the breech bolt and barrel together when in the forward position. The firing pin is positioned by the firing pin stop pin passing laterally through the bolt and a groove in the firing pin. In the under side of the breech bolt the locking block latch is pivoted and actuated by a coil spring positioned between the latch and the bolt. The latch blocks the lower end of the locking block from forward movement until levered up by the link as the rear end of the link rises at the end of the forward movement. The operating slide, by which the breech bolt is manually retracted to load and cock the gun, is positioned in a groove in the right side of the bolt and reciprocates with it when the gun functions. The slide also functions with the link and carrier dog to unlock the breech bolt and barrel.

(10) The trigger plate group (fig. 103) is composed of the hammer, trigger, safety sear, mainspring, and trigger safety together with their components. The hammer is pivoted in the forward end of the trigger plate and is functioned by the mainspring screwed to the tang of the trigger plate, and bearing on a roller in the lower end of the hammer. The mainspring is of the leaf type and is slotted to allow the upper end of the trigger, which acts as the sear, to protrude through it to engage the hammer. A lug extends from the upper rear face of the hammer in which there are two notches, one facing forward and the other to the rear. The trigger is pivoted in the rear end of the trigger plate; the lower end extending downward into the guard bow. The upper end of the trigger is U-shaped; the upper ends of the U, forming hooks facing towards each other. When in the cocked position, the notched hammer lug lies between these hooks and is held by one or the other as explained hereafter. This feature prevents the gun from firing automatically as the trigger must be released and again pulled before the gun will fire. The safety sear is pivoted on a lug on the trigger plate directly over the trigger and acts as a block to the trigger during the functioning of the operating mechanism. The safety sear is spring-actuated by a spring follower seated in the lug. The trigger safety is in form of a cylinder positioned in the trigger plate to the rear of the trigger and operating laterally in the trigger plate to block or free the trigger for pulling, to fire the gun.

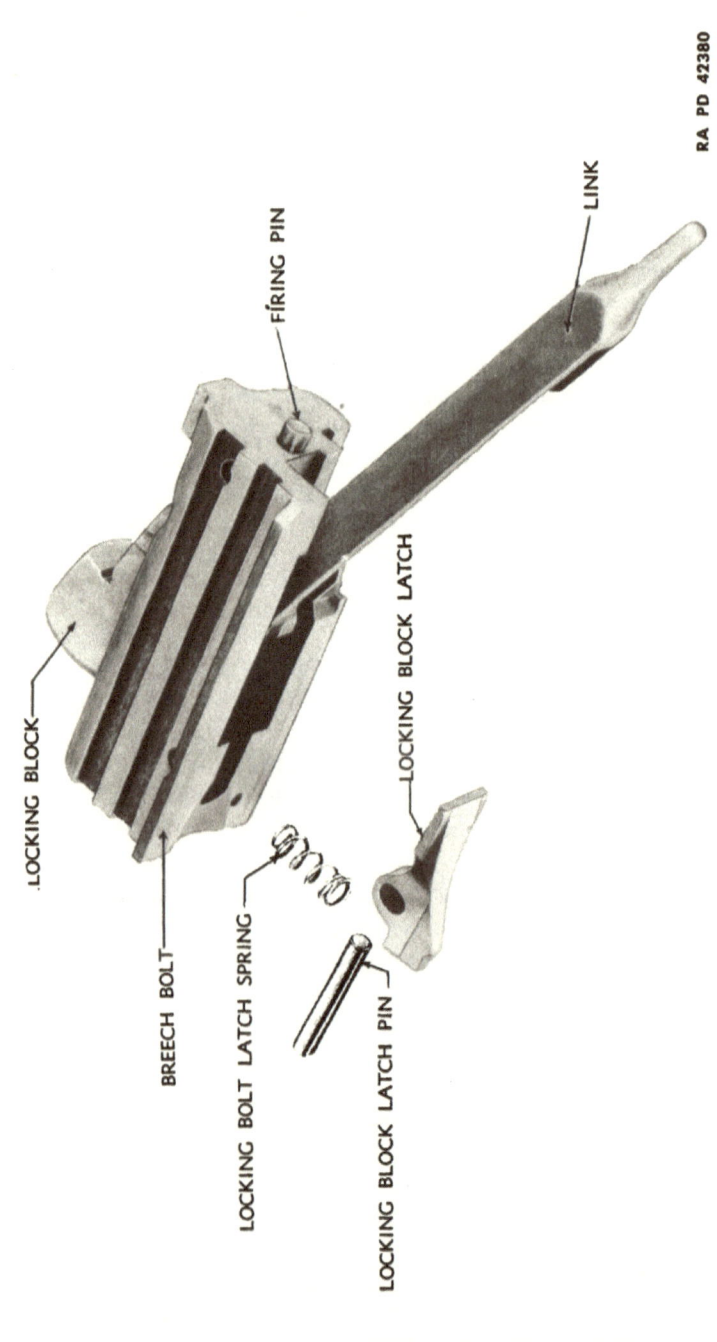

Figure 102—Breech Bolt Group—Rear, Bottom and Left Side View—Locking Block Latch Disassembled—Remington Shotgun M11—Sportsman

REMINGTON SHOTGUN, 12-GAGE, M11 AND SPORTSMAN

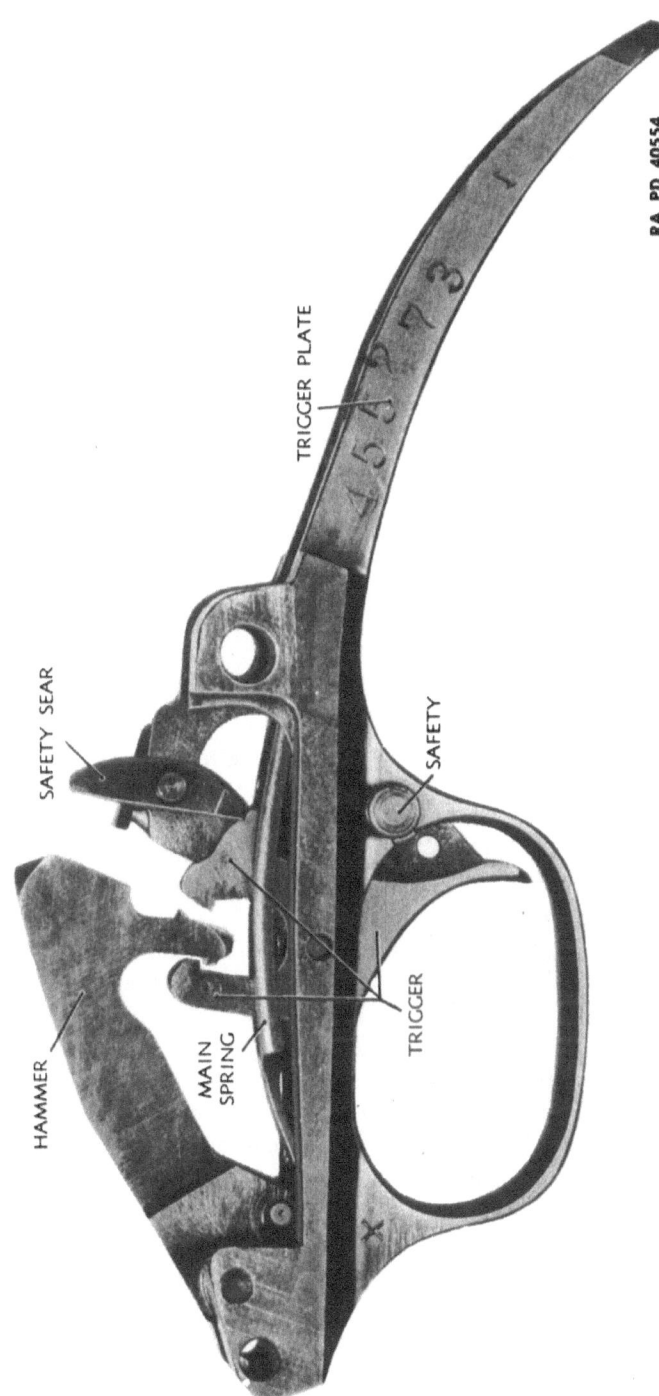

Figure 103—Trigger Plate Group—Left Side View—Remington Shotgun M11—Sportsman

SHOTGUNS, ALL TYPES

53. DATA.

Gage of bore	12
Boring of barrel—riot type	Cylinder
Boring of barrel—sporting skeet type	Improved cylinder
Boring of barrel—sporting trap type	Full choke
Type of action	Semiautomatic
Type of firing mechanism	Hammerless
Type of magazine	Tubular
Capacity of magazine—M11	4 shells
Capacity of magazine—Sportsman	2 shells
Length of barrel—riot type	20 in.
Length of barrel—sporting skeet type	26 in.
Length of barrel—sporting trap type	30 in.
Length of stock, receiver and fore end assembled (approx.)	31¾ in.
Length of assembled gun—riot type (approx.)	40 in.
Length of assembled gun—sporting skeet type (approx.)	46 in.
Length of assembled gun—sporting trap type (approx.)	50 in.
Weight of assembled gun—riot type (approx.)	7¾ lb
Weight of assembled gun—sporting skeet type—Sportsman (approx.)	8½ lb
Weight of assembled gun—sporting trap type (approx.)	8½ lb

54. OPERATION.

a. This gun is of the semiautomatic type as already explained, and is loaded and cocked by the force of recoil after the first shot. The only action necessary thereafter to fire the gun is to pull the trigger. The first shell, however, must be loaded into the chamber and the hammer cocked by manual operation. This is accomplished by retraction of the breech bolt by means of the operating slide (fig. 106).

CAUTION: During operation, the muzzle of the gun should always be pointing at a safe spot and the finger kept outside of the trigger plate guard bow.

b. When the gun is operated as a repeater, the shells are loaded into the magazine, and a shell then transferred to the chamber and the hammer cocked by manual retraction of the slide as already stated. When operating the gun as a single loader, the bolt can be retracted and hung in the rearward (open) position by retraction of the operating slide. This operation unlocks the bolt from the barrel, pulls it to the rear, and engages it with the dog of the carrier to hang it in the rearward position. The bolt is again released to move forward by pressing the carrier latch button, thus allowing the carrier to rise and the dog to release the bolt. The bolt springs forward and pushes the shell into the chamber of the barrel, and bolt and barrel are locked together as already explained.

c. The double hook on the trigger retains the hammer in the cocked

REMINGTON SHOTGUN, 12-GAGE, M11 AND SPORTSMAN

position whether the trigger is released betwen shots or not. The trigger, however, must be released and again pulled before the hammer is released from the cocked position to fire the gun.

d. The safety sear blocks the trigger from being pulled during the rearward and forward movement of the operating mechanism. The safety sear is engaged with the trigger to block its being pulled by the spring action of the safety sear follower, and disengaged from the trigger by the rear end of the link as the link locks the bolt and barrel together at the end of the forward movement. This feature prevents premature firing of the gun before the bolt and barrel are locked together in the forward position.

e. The trigger safety is positioned in the rear of the guard bow and operates laterally in the trigger plate (fig. 104). When pushed fully to the left, the trigger is free to be pulled and the gun fired. When pushed fully to the right, the trigger is blocked and cannot be pulled nor the gun fired. With a shell in the barrel chamber, it is best to block the trigger by pushing the safety to the right, as explained above, unless the gun is to be fired immediately.

f. **To Load the Magazine** (fig. 105).

(1) With breech bolt locked, press the carrier latch button fully inward to release the forward end of the carrier. Press shell nose down against carrier until it can be pushed forward against the magazine follower and into the rear of the magazine. As the base of the shell passes the end of carrier latch, pressure on latch button should be released slightly. Push the shell forward into the magazine until the base of the shell passes and is retained by the end of the carrier latch.

(2) Holding the carrier latch button depressed with the thumb, load another shell in the same manner pressing it against the base of the first shell loaded, and push both shells forward into the magazine until the second shell is caught and held as above. When the carrier latch button is released, the rearmost shell will spring back and be held by the locking block latch which projects from the bottom of the bolt.

NOTE: The magazine of the Sportsman Model Gun holds two shells, while that of the standard M11 Gun holds four shells.

g. **To Unload the Magazine.** Set safety at safe by pushing it all the way to the right. Then depressing carrier latch button, press carrier against breech bolt, and holding carrier thus, pull operating handle part way to the rear to release the shell from the magazine onto carrier, and lift shell from receiver. The shell in the barrel (if loaded) will return to the barrel when the slide is released.

h. **To Load the Barrel from the Magazine.**

(1) With the magazine loaded set safety at safe by pushing all the

TM 9-285
54

SHOTGUNS, ALL TYPES

Figure 104 — Operating Safety — Remington Shotgun M11 — Sportsman

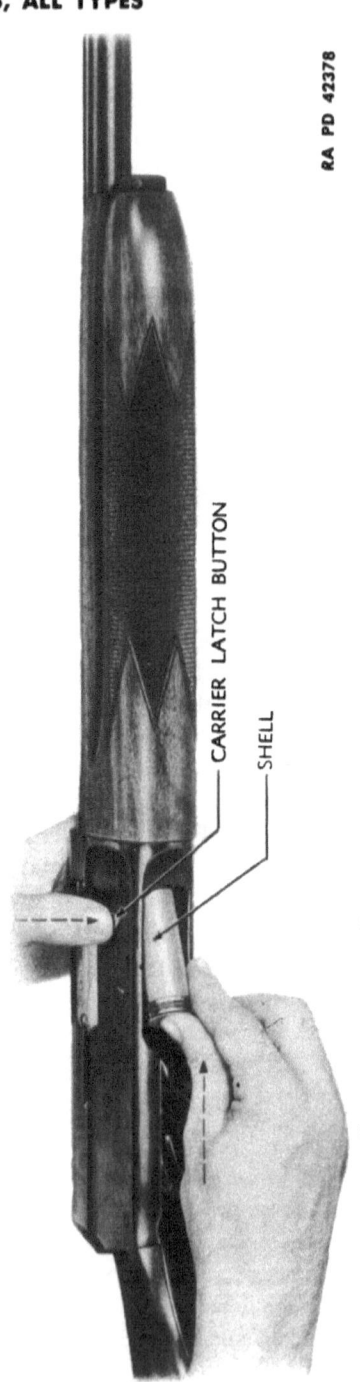

Figure 105 — Loading Magazine — Carrier Latch Button Depressed — Remington Shotgun M11 — Sportsman

TM 9-285
54

REMINGTON SHOTGUN, 12-GAGE, M11 AND SPORTSMAN

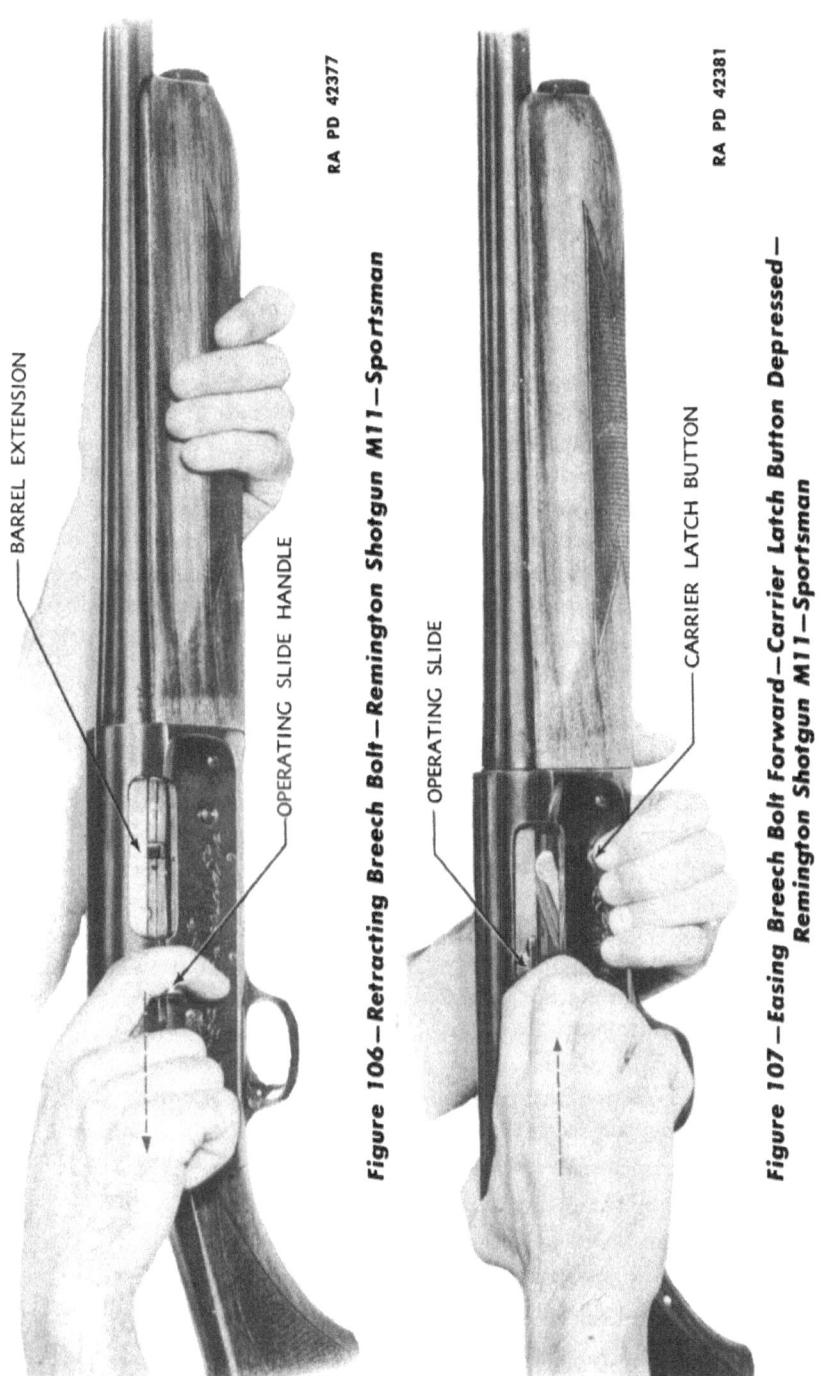

Figure 106 — Retracting Breech Bolt — Remington Shotgun M11 — Sportsman

Figure 107 — Easing Breech Bolt Forward — Carrier Latch Button Depressed — Remington Shotgun M11 — Sportsman

SHOTGUNS, ALL TYPES

way to the right. Then pull the operating slide fully to the rear and release. This action will cock the hammer and feed a shell from the magazine into the receiver. When the breech bolt springs forward, it will push the shell into the barrel and close and lock the action.

CAUTION: The fingers should be kept clear of the ejection opening in the receiver as well as the path of the operating slide handle during this operation to prevent injury to fingers due to the closing of the action.

(2) After a shell has been loaded from the magazine into the barrel, another shell may be loaded into the magazine (par. 54 f).

i. To Load the Barrel Only. With the magazine empty, set safety at safe by pushing all the way to the right and pull operating slide to rear and hang the bolt. Then place a shell, nose forward, into ejection opening in the right side of the receiver and allow it to settle on the carrier. *With fingers clear of ejection opening and path of the operating slide handle, press carrier latch button.* The breech bolt will thus be released to spring forward; move the shell into the barrel and close and lock the gun. The barrel is now loaded and the hammer cocked. With the barrel loaded, it is best to allow the safety to remain at safe, to block the trigger, unless the gun is to be fired immediately.

j. To Unload Barrel Without Unloading Magazine. If it is necessary to unload the barrel without unloading the magazine, it can be accomplished as follows: Set safety at safe by pushing all the way to the right. Press carrier latch button and then push shell, which protrudes part way from the magazine, fully into the magazine until it is retained by the forward end of the carrier latch. Keep shell in magazine by continuing to press on latch button, and retract operating slide (fig. 106) to eject shell in barrel. Then, still pressing on carrier latch button, allow breech bolt to close slowly (fig. 107) watching to see that the bolt closes on an empty barrel chamber.

k. To Completely Unload the Gun. When gun is completely loaded (barrel chamber and magazine) it should be unloaded as follows:

(1) Set safety at safe by pushing all the way to the right.

(2) Unload magazine first as described in g above.

(3) With magazine empty unload the barrel chamber by pulling operating slide fully to rear, thus ejecting the shell in the barrel.

(4) Inspect barrel and magazine to make sure gun is fully unloaded.

(5) When gun is fully unloaded, the action may be closed by holding operating handle, pressing carrier latch button, and easing breech bolt to the forward position.

TM 9-285

REMINGTON SHOTGUN, 12-GAGE, M11 AND SPORTSMAN

55. FUNCTIONING.

a. As already explained, the functioning of this semiautomatic gun is accomplished through the force of recoil produced by the expanding powder gas acting through the shell against the breech bolt when the gun is fired. In this explanation of functioning, it is assumed that the gun is fully loaded and has just been fired.

b. As the powder gas expands, it forces the breech bolt to the rear along with the barrel which is locked to it by the locking block. As the bolt moves rearward, it pushes the link, which is pivoted to the rear end of the locking block, to the rear against the force of the expanded action spring with which it is engaged. The barrel is, at the same time, pulled to the rear by the breech bolt against the force of the expanded recoil spring and the resistance of the friction ring and piece, mounted on the magazine tube, acting against the rear face of the barrel guide attached to the barrel. Thus the action spring and recoil spring are compressed during the rearward movement by the bolt and barrel respectively. The resistance to compression of these springs and the friction set up by the contracted friction piece, act together against the force of the expanding powder gas upon the breech bolt (par. 58).

c. The breech bolt and barrel move to the rear, locked together, to the end of the rearward movement. This point is roughly determined by the opposing force of the springs and friction piece as explained above which stops the groups and then starts them forward. A fibre cushion riveted to the inner rear face of the receiver cushions the rear end of the bolt and prevents it from striking the receiver if it recoils too far on the rearward movement. As the breech bolt and barrel start forward, the dog on the rear of the carrier, which has been pressing upon the lower face of the bolt through the pressure of the carrier plunger spring, engages with the operating slide and checks the slide. The forward end of the carrier is prevented from rotating upward by the carrier latch. The operating slide, which has been bearing on the forward face of the link, likewise checks the link. The bolt, continuing forward, unlocks itself from the barrel as the locking block is rotated downward out of the locking aperture in the barrel by the momentarily held link pulling upon the locking block. As the bolt is unlocked, the barrel springs forward, pulled by the force of the now compressed recoil spring bearing on the rear face of the barrel guide.

d. The bolt is momentarily held and released as follows:

(1) When the bolt is unlocked, it is held to the rear by the engagement of the carrier dog with the operating slide and link as already explained. The bolt is held thus until the forward end of the carrier, which has been held in the lower position by the carrier latch, is freed to be rotated upward by the force of the action spring acting through

SHOTGUNS, ALL TYPES

the link and slide which are bearing upon the carrier dog. As the forward end of the carrier rises, the dog is disengaged from the slide and the bolt is free to move forward.

(2) The carrier latch is disengaged from the carrier by the shell which has been fed from the magazine into the receiver. This shell is held by the shell stop and released by the barrel as it springs forward when unlocked as explained in f below. The released shell springs to the rear and strikes and depresses the rear end of the carrier latch, and thus disengages the latch from the carrier. The carrier is thus freed to be rotated upward as explained above.

(3) When the last shell has been fired, the bolt is held to the rear by the carrier dog as explained. There being no shell in the receiver to release the carrier by striking and depressing the carrier latch, the bolt will be held until the carrier is released manually by pressure on the carrier latch button.

(4) The bolt thus released as explained above moves forward, propelled by the expanding action spring by means of the link and locking block. As the bolt is pushed forward, the link is at an angle to the axis of the bolt. In this position, the forward end of the link lies above the rear end of the locking block latch which is pivoted in the bolt and bearing on a shoulder on the locking block, preventing its rotation upward. The link thus pushes the bolt forward although the link is pivoted to the lower end of the locking block. As the bolt thus propelled nears the forward position, the rear end of the link rises until the link is nearly parallel to the axis of the bolt. As the rear end of the link rises, the forward end levers the locking block latch down from engagement with the locking block, and immediately begins to rotate the locking block upward in the bolt. This movement is so timed that the locking block is rotated fully through the bolt and into the locking aperture in the barrel extension as the bolt reaches the forward position, thus locking bolt and barrel together again. The rear lug on the locking block, striking the sloping rear end of the barrel extension, cushions the bolt and starts the locking block rotating upward.

e. When the bolt moves rearward, pulling the barrel, on the recoil movement, the fired shell, held in position on the face of the bolt by the extractor on the right-hand side, moves to the rear with the bolt and barrel. As the barrel is freed from the bolt at the beginning of the forward (counterrecoil) movement, it springs forward as explained. As the rear of the barrel extension passes the face of the bolt, the ejector pinned in the left wall of the barrel extension strikes the base of the fired shell held to the face of the bolt by the extractor and knocks it to the right out of the ejection opening in the receiver, pivoting it about the extractor.

REMINGTON SHOTGUN, 12-GAGE, M11 AND SPORTSMAN

f. As the barrel extension continues forward, it strikes and disengages the shell stop which has been holding the next shell to be loaded. The shell springs rearward, propelled by the force of the magazine spring, to release the carrier latch and hence the bolt, and is then lifted to the bore line by the carrier which is rotated up by the link as explained in d above. The shell is pushed forward into the barrel by the bolt as it moves forward and the carrier spring returns the carrier to the lower position.

g. As the rear end of the carrier latch is depressed by the shell and thus disengaged from the carrier, the forward end is rotated outward into the path of the next shell in the magazine. The latch holds the shell thus until the carrier moves downward again after having lifted the previous shell to the bore line. As the carrier moves downward it releases the rear end of the carrier latch which it has been holding depressed. As the rear end springs outward, the forward end rotates in towards the receiver wall and thus frees the shell it has been holding. This shell springs to the rear and is held by the forward end of the locking block latch on the bolt, which has now reached the forward position, as the carrier reaches the lower position and thus retains the shell in the receiver. The above functioning prevents double feeding. The shell is held thus by the locking block latch until the bolt starts rearward again, when the shell follows it until caught and retained by the shell stop, from which it is subsequently released by the barrel extension on its forward movement as already explained.

h. The hammer is cocked by the link, which straddles it, on the rearward movement of the bolt. The sear hooks on the trigger (fig. 103) engage with notches in the hammer to hold it in the cocked position, whether the trigger has been released between shots or not. The safety sear blocks the trigger preventing its being pulled until disengaged by the link as the bolt reaches the locked position. Detailed explanation of the functioning of the trigger is as follows:

(1) The upper part of the trigger is shaped like a U with two sear hooks facing each other. A lug projecting from the rear face of the hammer is notched on both sides to engage with the sear hooks on the trigger. The notched lug of the hammer rotates down between these hooks when cammed back by the link on the rearward movement of the bolt. If the trigger is in the forward (released) position, the forward hook engages with the forward notch in the hammer. If the trigger is in the rearward (pulled) position, the rear hook engages with the rear notch in the hammer. Thus the hammer is caught and held in the cocked position whether the trigger has been released or not. If the trigger has been held in the pulled position and the hammer caught by the rear hook, the trigger must be released and again pulled before the gun can be fired. Release

SHOTGUNS, ALL TYPES

of the trigger disengages the rear hook but the hammer is immediately caught and held by the forward hook. When the trigger is again pulled, the hammer is released and rotates forward, propelled by the force of the main spring, to strike the rear of the firing pin positioned in the breech bolt.

(2) The safety sear, however, has been engaged with the rear of the trigger by the action of its spring follower. In order to pull the trigger, the safety sear must be disengaged. This is accomplished by the rear end of the link, which straddles the safety sear as it reaches the forward position to lock the bolt. This feature prevents the gun from being prematurely fired before the breech bolt is locked to the barrel.

i. The trigger safety functions laterally in the trigger plate just behind the trigger. A web on the rear face of the lower part of the trigger slips into a notch in the safety when the safety is pushed to the left and thus disengaged, allowing the trigger to be pulled. When the safety is shifted laterally to the right, the trigger web cannot enter the notch in the safety and therefore cannot be pulled.

56. REMOVAL OF GROUPS (figs. 101, 108, and 109).

a. Groups and parts should be removed and replaced in the order given below. Groups and parts when removed should be placed on a clean, flat surface and care exercised to prevent loss of screws and small parts. Remove as follows:

(1) BARREL AND FORE END.

(a) With breech bolt in the forward position and gun held in vertical position, rest butt on solid surface.

(b) Push barrel back into receiver a short distance and hold in that position to relieve pressure of recoil spring on fore end; unscrew takedown screw (Sportsman) or magazine cap (M11), located at forward end of fore end from magazine tube (fig. 124). If screw is tight, breech bolt may be retracted and hung in rearward position and fore end held down against force of recoil spring while loosening screw. Push out cross pin in screw head and use as lever (fig. 110).

(c) When take-down screw (Sportsman) or magazine cap (M11) is fully disengaged from magazine, ease barrel upward out of receiver until free of pressure of recoil spring (fig. 125).

(d) Slide fore end, with take-down screw attached, up and off magazine and barrel and then pull barrel up and out of receiver (fig. 111).

(e) Note relative positions of friction piece, friction ring, and recoil spring as mounted on the magazine tube and if removed, replace in identical manner. Refer to paragraph 58 for assembly of these parts for varying recoil pressures.

TM 9-285
56

REMINGTON SHOTGUN, 12-GAGE, M11 AND SPORTSMAN

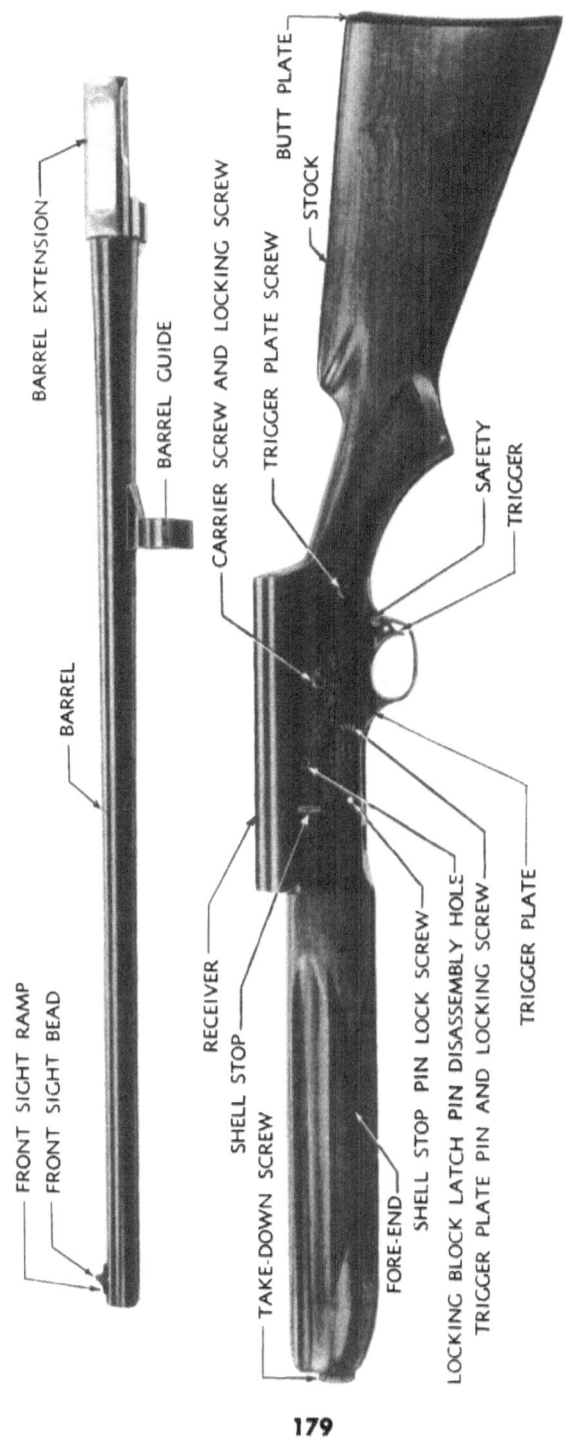

Figure 108—Gun Taken Down—Left Side View—Showing Location of Parts - Remington Shotgun, Sportsman

TM 9-285

SHOTGUNS, ALL TYPES

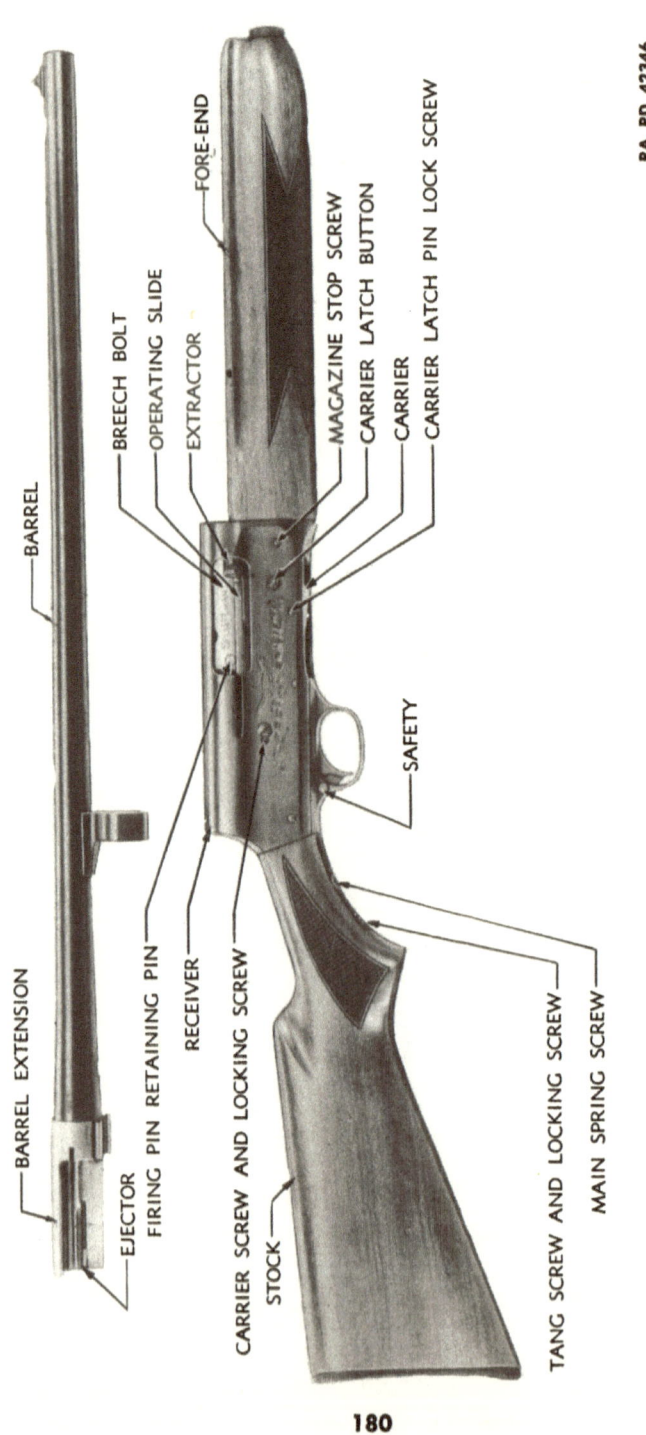

Figure 109—Gun Taken Down—Right Side View—Showing Location of Parts—Remington Shotgun, Sportsman

TM 9-285

REMINGTON SHOTGUN, 12-GAGE, M11 AND SPORTSMAN

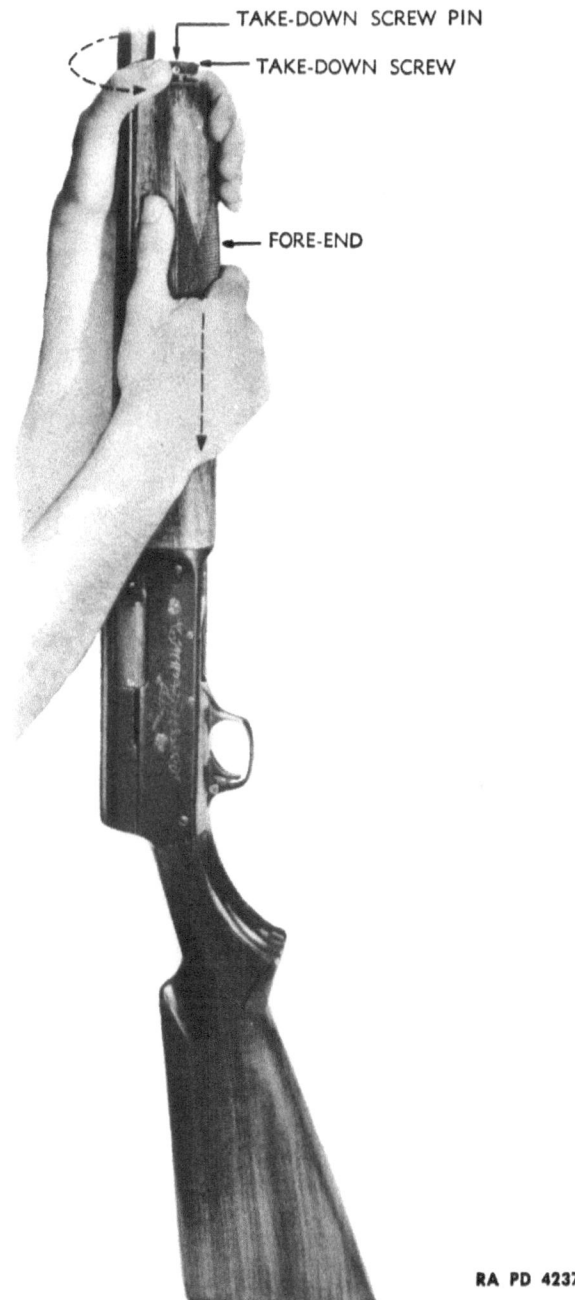

Figure 110—Disengaging Take-down Screw While Holding Fore End Against Spring Tension—Remington Shotgun, Sportsman

SHOTGUNS, ALL TYPES

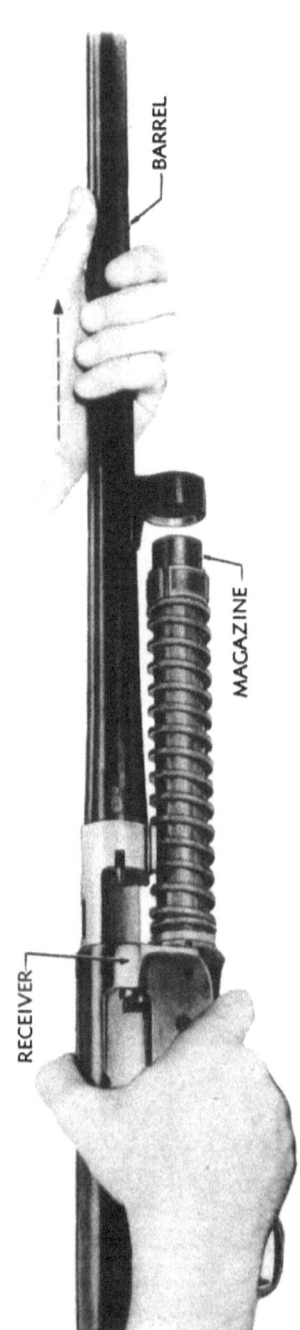

Figure 111—Removing Barrel from Magazine and Receiver—Remington Shotgun M11—Sportsman

REMINGTON SHOTGUN, 12-GAGE, M11 AND SPORTSMAN

(f) Replace fore end on magazine tube, press down against recoil spring, and screw in take-down screw or magazine cap (M11) to hold recoil spring, friction piece, and friction ring in position. Exercise care not to lose friction piece spring.

CAUTION: Do not press carrier latch button allowing breech bolt to spring forward after barrel is removed from receiver. To move bolt forward and remove tension from action spring, grasp operating slide handle firmly; then, press carrier latch button and ease bolt to forward position against force of action spring.

(2) BUTT STOCK. Due to expansion of action spring when breech bolt is removed, it is necessary to remove this spring before removing the groups from the receiver. To remove the spring, the butt stock must first be removed. This is accomplished by removing tang screw locking screw, and then the tang screw from under side of trigger plate tang (rear screw in tang), and pulling butt stock to rear from receiver and action spring tube. (If withdrawal of butt stock is difficult, remove trigger plate pin, press carrier latch button to release carrier, and press forward end of trigger plate upward slightly into receiver.)

(3) TRIGGER PLATE GROUP.
(a) With butt stock removed, remove trigger plate pin locking screw from lower left hand side of receiver (near center) and, using pin drift, drive out trigger plate pin from right to left.

(b) Remove trigger plate screw from lower rear corner of left side of receiver, and pull trigger plate group downward out of receiver.

(4) CARRIER.
(a) Press forward end of carrier spring from under retaining stud and remove spring from pivot stud. Be careful spring does not fly out when disengaging forward end.

(b) Remove locking screws and then, carrier screws from right and left sides of receiver and lift out carrier. Carrier screws are largest screws in sides of receiver, just to rear of center.

(5) ACTION SPRING.
(a) With butt stock, trigger plate and carrier removed, press in the wooden plug at rear end of action spring tube, and using pin drift, push out cross pin.

(b) Holding plug with finger against force of spring, withdraw drift slowly and ease plug and spring to rear out of tube. When fully expanded, pull plug, spring, and spring follower to rear out of tube.

(6) BREECH BOLT AND LINK. Recoil spring, friciton piece, and ring must be moved forward on the magazine tube before breech bolt can be removed from receiver.

TM 9-285
56-57

SHOTGUNS, ALL TYPES

(a) Move breech bolt until locking block latch pin alines with hole in left side of receiver to rear of shell stop. Then, using straight pin drift, push pin clear through breech bolt and receiver from right to left; then, withdraw pin drift slowly at same time holding down locking block latch against force of latch spring. When drift is removed, ease latch upward and remove latch and spring.

(b) Swing rear end of link downward to clear operating slide, and move breech bolt forward out of receiver with the link attached. The operating slide will slide out of the breech bolt as it moves forward and the handle of the slide strikes the forward edge of the ejection opening in the receiver.

57. REPLACEMENT OF GROUPS.

a. Groups and parts should be thoroughly cleaned, lightly oiled, and lubricated, if necessary, before replacing. Replace as follows:

(1) BREECH BOLT AND LINK.

(a) With recoil spring, friction piece, and ring moved forward on the magazine tube, slide the link and breech bolt, link first, into the forward end of the receiver, so that guides on lower face of breech bolt mate with guideways in receiver.

(b) Move breech bolt to rear into receiver until half the bolt has entered. Then, holding operating slide by the handle, slip rear end into receiver to rear of bolt, through ejection opening in right side. Hold slide parallel to bolt and move bolt to rear, at the same time mating guides on slide with guideways in right side of bolt. As slide enters bolt, move bolt to rear thus seating slide in bolt. The link must be pointed downward to accomplish the mating.

(c) Slide recoil spring, friction piece, and friction ring to rear on magazine tube, in their proper order and position and replace fore end to hold in place (par. 56 a (1) *(e)* and *(f)*).

(d) Slide breech bolt to rear until locking block latch pin holes in bolt and receiver aline, insert latch pin through left side of receiver and left wall of bolt to hold in position, then, replace latch spring in seat in bolt and place latch on top of spring with long flat end to rear and drilled lug down. Press latch down on spring, aline hole in latch with holes in receiver, and bolt and push pin through latch and bolt until flush with both sides of bolt. Move bolt to see that pin does not interfere with receiver.

(2) CARRIER.

(a) With receiver bottom side up, move breech bolt forward and push link towards top of receiver.

(b) With carrier dog up and to rear, slide carrier, beneath spring retaining stud, into receiver and aline screw holes in carrier with those in

REMINGTON SHOTGUN, 12-GAGE, M11 AND SPORTSMAN

receiver. Replace carrier screws; then, turn screws until cuts in screw heads aline with countersinks for locking screws and replace locking screws. (Catch threads in both carrier screws before tightening either screw. Be careful not to cross threads).

(c) Slide carrier spring onto the pivot stud in left rear of receiver so that short leaf rests on rear of carrier. Then, press down long leaf of spring and slip under retaining stud head just forward of pivot stud. Hold spring to prevent slipping from pivot stud while positioning long leaf.

(3) ACTION SPRING.

(a) With breech bolt forward, insert action spring follower first (assembled to spring) into rear end of action spring tube. Push through tube and mate rear nose of link with indent in follower.

(b) Press action spring into tube until head of wooden plug (assembled to spring) is flush with end of tube. Turn plug until pin hole in plug alines with that in tube and insert plug pin. Still holding plug, push pin through until flush with tube. A pin drift may be used to hold plug while inserting pin.

(4) TRIGGER PLATE GROUP.

(a) With the breech bolt forward and the hammer cocked, press trigger plate group, with tang to rear, horizontally upward into the rear under side of the receiver and adjust until pin and screw holes in trigger plate and receiver aline. If plate does not seat easily, retract operating slide slightly to allow safety sear to enter slot in link.

(b) Insert trigger plate pin from left side and push through until head is flush with face of receiver.

(c) Using screwdriver in slotted head of pin, turn pin until cut in pin head alines with locking screw countersink in receiver. Then replace locking screw.

(d) Insert trigger plate screw from left side of receiver, catch threads, and turn down tightly.

(5) BUTT STOCK.

(a) Push butt stock on over action spring tube until stock fits snugly and evenly against rear face of receiver, and tang screw hole in trigger plate tang and stock aline. Stock may be seated by striking butt smartly with heel of hand. Do not strike butt on hard surface.

(b) Replace tang screw and screw in tightly until cut in screw head alines with locking screw countersink in tang; then replace locking screw. If replacement of butt stock is difficult, remove trigger plate pin. Replace pin after assembly.

(6) BARREL AND FORE END.

(a) With breech bolt forward, fore end removed from magazine tube, and recoil spring, friction piece, and friction ring in their proper order

SHOTGUNS, ALL TYPES

and position, hold gun in vertical position and rest butt of stock on firm surface.

(b) Slide the barrel extension into open end of receiver so that guides on barrel extension mate with guideways in receiver and barrel guide slips over magazine tube and bears on friction assembly on tube.

(c) Slide fore end on barrel and magazine tube as far as it will go.

(d) Grasp muzzle end of barrel and push barrel down into receiver against spring force until the barrel extension is wholly within the receiver. Then, press fore end down until it mates evenly with receiver and engage and screw down take-down screw (Sportsman) or magazine cap (M11) until tight. Pin in screw can be pushed out and used as a lever, if necessary. Be sure fore end is fully mated flush with receiver and take-down screw or magazine cap (M11) is tight.

(e) Release barrel and operate gun to test assembly.

58. USE OF FRICTION RING.

a. Proper use of friction ring and piece will reduce recoil and prevent excessive wear of parts. To change friction adjustment, remove fore end and barrel and assemble the recoil spring, friction piece (with friction spring assembled), and the friction ring as prescribed below.

(1) FOR HEAVY LOADS. When heavy loads are used, ranging from $3\frac{1}{4}$ to $3\frac{3}{4}$ drams (or equivalent) of powder, assemble as follows:

(a) Place the recoil spring on the magazine tube first so that it bears directly against the receiver. The Sportsman Model Gun has a stop ring on the magazine next to the receiver, held in position by a set screw. This ring should not be removed.

(b) Place the friction ring next on the magazine tube with the outside bevel to the rear against the recoil spring.

(c) Place the friction piece (with friction spring assembled) on the magazine tube ahead of the friction ring.

NOTE: Refer to figure 112, position 1.

(2) FOR LIGHT LOADS. When light loads are used, 3 drams (or equivalent) or less of powder, assemble as follows:

(a) Place friction ring next to receiver with outside bevel facing forward, away from receiver.

(b) Place recoil spring next to friction ring.

(c) Place friction piece (with friction spring assembled) next to recoil spring.

NOTE: Refer to figure 112, position 2.

(3) FOR USE WITH CUTTS COMPENSATOR. When the Cutts Compensator is mounted to the gun barrel, assemble as follows:

TM 9-285

REMINGTON SHOTGUN, 12-GAGE, M11 AND SPORTSMAN

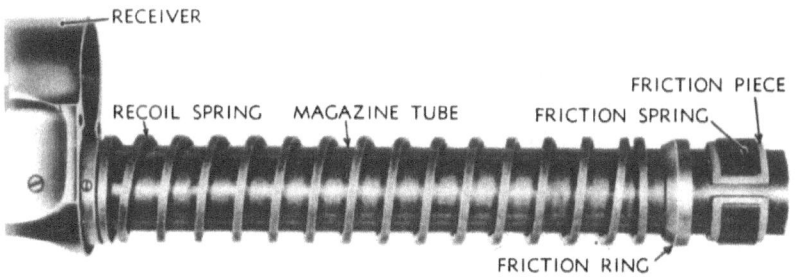

Sportsman—POSITION 1—Heavy Loads

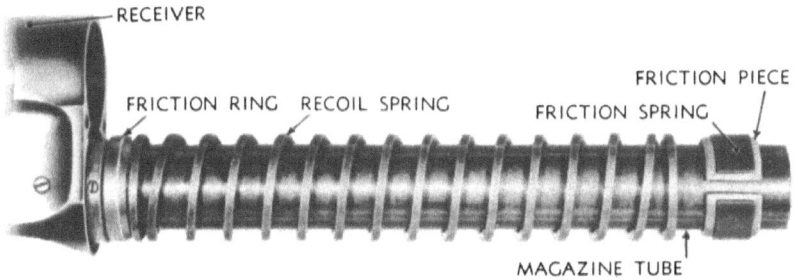

Sportsman—POSITION 2—Light Loads

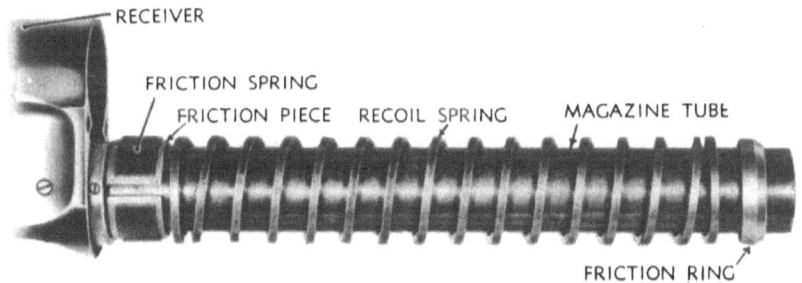

Sportsman—POSITION 3—With Cutts Compensator

Figure 112—Recoil Adjustments—Remington Shotgun M11

SHOTGUNS, ALL TYPES

(a) Place friction piece (with friction spring assembled) next to receiver.

(b) Place recoil spring next to friction piece.

(c) Place friction ring next to recoil spring with outside bevel facing forward, away from spring.

NOTE: Refer to figure 112, position 3. The Cutts Compensator is a slotted tube, sometimes attached to the muzzle end of the barrel for the purpose of reducing recoil and controlling shot pattern.

(4) The friction is progressively reduced (subpar. *a* to *c* above), figure 112, positions 1 to 3. In order to reduce recoil and excessive wear of parts, use greatest friction possible which will give satisfactory functioning of gun. If gun fails to eject, however, try next lighter friction adjustment; if failures occur with position 1, try position 2; if failures occur with position 2, try position 3.

(5) Magazine tube should be kept free from foreign matter and rust and lightly oiled. If gun does not function smoothly and freely, lubricate friction piece and ring lightly.

59. FIELD INSPECTION.

a. With the gun completely assembled, test the mechanism for proper functioning. Fired shells may often be used for testing, when dummy shells are not available, by turning in the uncrimped end so that the length of the shell will approximate that of a live shell. Use of live shells for testing is prohibited.

CAUTION: Keep fingers clear of ejection opening and path of operating slide handle as bolt moves forward with considerable force when released and may cause injury to operator. Be sure gun is fully unloaded before inspection.

b. Operate the gun as follows:

(1) Grasping fore end with left hand and, operating slide handle with right hand, pull breech bolt all the way to the rear until bolt is caught and hung by carrier dog. Barrel should remain in forward position.

(2) With bolt hung in rearward position, press carrier latch release button and allow the bolt to spring forward. The bolt should spring forward smartly to the extreme forward position to lock with the barrel.

(3) Load a dummy shell into the magazine (par. 54 f). Retract breech bolt fully by means of operating slide and attempt to hang bolt in rearward position. Bolt should not hang, due to shell bearing on carrier latch (par. 55 d).

REMINGTON SHOTGUN, 12-GAGE, M11 AND SPORTSMAN

(4) Allow bolt to spring forward to load the shell into the barrel chamber. Retract bolt fully by means of operating slide to eject the shell and attempt to hang bolt. Bolt should hang, as there is no longer a shell in the receiver to bear upon the carrier latch.

(5) Test gun for doubling as follows:

(a) With gun fully unloaded, pull bolt fully to the rear by means of the operating slide handle until it is hung by the carrier dog.

(b) Hold bolt thus with right hand, and with the left hand press carrier latch button to release the bolt, and ease bolt forward just enough to release it from the hung position (about ¼ in.). Then, with left hand, pull the trigger.

(c) Holding trigger retracted, ease bolt slowly forward to the locked position.

(d) Then ease pull on trigger slowly until entirely removed and the trigger moves forward. The hammer should not release to fire the gun during the slow release of the trigger.

(e) When trigger is fully released, pull it again. The hammer should release to fire the gun when the trigger is thus pulled.

NOTE: This test is to insure that the gun will not double (fire automatically) if the trigger is not released (c (10) below). To insure the gun will not double, it is desirable that the movement of the trigger at the center point of finger contact be not less than 1/16 inch.

(6) Retract the bolt fully and, then, holding operating slide, press carrier latch button and allow the bolt to close slowly. Release and attempt to pull the trigger at intervals during the closing of the bolt. The trigger should not pull to fire the gun until the operating slide handle is within 1/16 to 1/8 inch from the normal forward position. This test is to make sure that the gun will not fire until the locking block is engaged with the recoil shoulder of the barrel extension and bolt and barrel thus locked together.

(7) Retract bolt fully and then allow it to move fully forward to the locked position, thereby cocking the hammer. Then, push the trigger safety all the way to the right and attempt to pull the trigger. The trigger should not pull nor the hammer be released to fire the gun.

(8) Push the safety all the way to the left and attempt to pull the trigger. The trigger should pull and the hammer be released.

c. When the gun does not operate and function smoothly when tested as above, damaged or improperly assembled parts are indicated as follows:

(1) BOLT DOES NOT HANG IN REARWARD POSITION WITH GUN UNLOADED. May be due to broken or worn carrier dog, worn retaining

SHOTGUNS, ALL TYPES

notch of operating slide or link shoulder, broken or missing carrier latch spring, broken carrier latch or foreign matter under forward end of latch.

(2) BOLT DOES NOT SPRING FORWARD SMARTLY WHEN RELEASED. May be due to broken action spring, link broken or improperly seated in action spring follower, jammed or broken carrier or carrier latch, broken locking block latch or broken or missing latch spring.

(3) BOLT DOES NOT LOCK TO BARREL. May be due to foreign matter in receiver or barrel extension, burred locking block or locking apertures in barrel extension, or broken or jammed locking block latch spring.

(4) CARRIER DOES NOT FUNCTION TO HANG BOLT OR LIFT SHELL. May be due to broken carrier dog or follower spring, jammed or broken carrier latch, foreign matter in mechanism, or worn or broken operating slide.

(5) CARRIER DOES NOT RETURN TO LOWER POSITION. May be due to broken or improperly assembled carrier spring, or jammed or broken carrier latch.

(6) SHELL DOES NOT SPRING FROM MAGAZINE WHEN RELEASED. May be due to broken or kinked magazine spring, dented tube, bent follower, or foreign matter or rust in tube.

(7) HAMMER DOES NOT COCK OR SLIPS. May be due to worn or broken hooks on trigger, or notches in hammer, broken trigger or mainspring, or foreign matter in mechanism.

(8) SAFETY SEAR DOES NOT RELEASE TRIGGER WHEN BOLT IS LOCKED. May be due to broken or jammed safety sear spring follower, or damaged safety sear.

(9) TRIGGER SAFETY DOES NOT OPERATE. May be due to burred web on rear of trigger, burs in slot in safety, broken, or bent leaf on trigger spring which bears on safety ball, or foreign matter in aperture.

(10) HAMMER IS RELEASED WHEN TESTED (par. 59 b (5)). May be due to worn or broken hooks on hammer or trigger, foreign matter in notches, or spread or broken upper U-end of trigger.

(11) THE FOLLOWING MALFUNCTIONS MAY OCCUR WHEN GUN IS FIRED:

(a) Barrel Is Not Unlocked From Breech Bolt. May be due to broken or damaged carrier dog or carrier latch, worn operating slide notch, or broken slide.

(b) Shells Do Not Eject. May be due to broken ejector in barrel extension, broken extractor or incorrect friction adjustment (par. 58 a (4)).

REMINGTON SHOTGUN, 12-GAGE, M11 AND SPORTSMAN

(c) Shell Stop Does Not Function. May be due to broken stop or broken or missing stop spring.

(d) Two Shells Enter Receiver at Once (Double Feeding). May be due to broken carrier latch or carrier dog.

d. Inspect barrel and test trigger pull (par. 3 *n*).

e. In addition to inspection of the gun for operation and functioning, the gun should be inspected generally for condition and defects noted. Attention should be directed to such defects as cracked wooden parts, cracked or deformed metal parts, dented or rusted magazine tube, loose screws and pins, loose or binding parts or assemblies, loose barrel or magazine, loose stock or butt plate, rust, dents, burs, or excessive wear of parts. If defects are such that early malfunction of the gun is indicated, the gun should be turned over to ordnance personnel for inspection and correction.

f. Where defects and malfunctions cannot be remedied by cleaning, lubrication and simple adjustments of assembly, which lie within the scope of using troops, the gun should be turned over to ordnance personnel for a thorough inspection, correction and/or repair.

g. Removal of burs on working parts, trigger adjustments and like corrections should not be attempted by using troops, as stoning of parts must be exacting, the angle of the faces concerned must not be changed, and volume of metal must not be materially reduced.

h. In addition to inspection of the barrel (d above), it should be inspected for looseness and alinement in the barrel extension. The draw marks on barrel and barrel extension should be in alinement as misalinement of these parts will cause binding of the barrel guide with the magazine tube and affect the functioning of the gun when assembled. If misalinement is evident, the gun should be turned over to ordance personnel for correction.

i. If shell appears unnecessarily loose in chamber with breech bolt locked, the gun should be turned over to ordnance personnel to be checked for headspace and worn locking block.

j. Adjustment and maintenance of the gun in the case of using troops is limited to the removal and replacement of the parts and groups of parts, as outlined in paragraphs 56 and 57, together with cleaning, lubrication, and such adjustments as are necessary in assembling the gun as outlined.

60. CLEANING AND LUBRICATION.

a. Cleaning, oiling, and lubrication may be accomplished in a manner

SHOTGUNS, ALL TYPES

similar to that described for the Winchester Gun M97 (par. 11). Attention should be given to corresponding parts and surfaces when lubricating.

b. The fore end and barrel should be removed from the receiver for cleaning the bore, and for thorough cleaning, the groups should be removed from the receiver and assemblies cleaned, oiled and lubricated as directed.

c. Points to lubricate are:

(1) Outer surface of barrel extension and barrel extension guides.

(2) Friction piece and friction ring when necessary, where they bear on magazine tube. Tube should be kept clean and lightly oiled. Too much oil will reduce friction, and prevent proper functioning of gun (par. 58 a (5)).

(3) Breech bolt guides and surface, occasionally.

(4) Link pin, connecting locking block with link.

(5) Locking block guides.

(6) Trigger and hammer pins.

(7) Carrier (trunnion) screws.

(8) Action spring follower. Spring should be removed occasionally and tube cleaned and oiled.

(9) Safety sear stud and spring follower.

d. Oiling and lubrication should be light, as excess oil collects foreign matter and powder residue which will become gummy, impede functioning of gun, and produce undue wear of parts. In very cold climates, oiling and lubrication should be reduced to a minimum. Only surfaces showing signs of wear should be lightly oiled. Refer to "Special Maintenance," section XI.

Section IX

SAVAGE SHOTGUN, 12-GAGE, M720

	Paragraph
Description	61
Data	62
Operation	63
Functioning	64
Removal of groups	65
Replacement of groups	66
Use of friction ring	67
Field inspection	68
Cleaning and lubrication	69

61. DESCRIPTION.

a. Identification marks on this gun are generally to be found as follows:

(1) Name of maker and model number of the gun are stamped on the left side of barrel near the rear end, and the chamber length and gage size in the corresponding position on the right side of the barrel.

(2) The word "SAVAGE" is embossed on the left side of the receiver.

(3) The serial number of the gun is stamped on the forward lower face of the receiver and the left side of the trigger guard tang.

(4) The letters U.S. are stamped on the forward left face of the receiver.

b. This gun (figs. 113 and 114) is an autoloading or semiautomatic shotgun similar in every way to the Remington Gun M11 covered in Section VIII, which varies slightly from the Sportsman Model. Like the Remington gun this gun is recoil operated, the force generated by the expanding powder gas acting against the breech bolt on the recoil movement, together with the force generated by the compressed recoil and action springs on the counterrecoil movement, operates the mechanism of the gun to load, feed, extract and eject the shell, and cock the hammer. The trigger must be pulled each time to fire the gun.

c. This gun is so constructed that the barrel can easily be removed from the receiver by removal of the magazine cap and pulling fore end and barrel from the receiver. This construction facilitates cleaning and transportation.

TM 9-285

SHOTGUNS, ALL TYPES

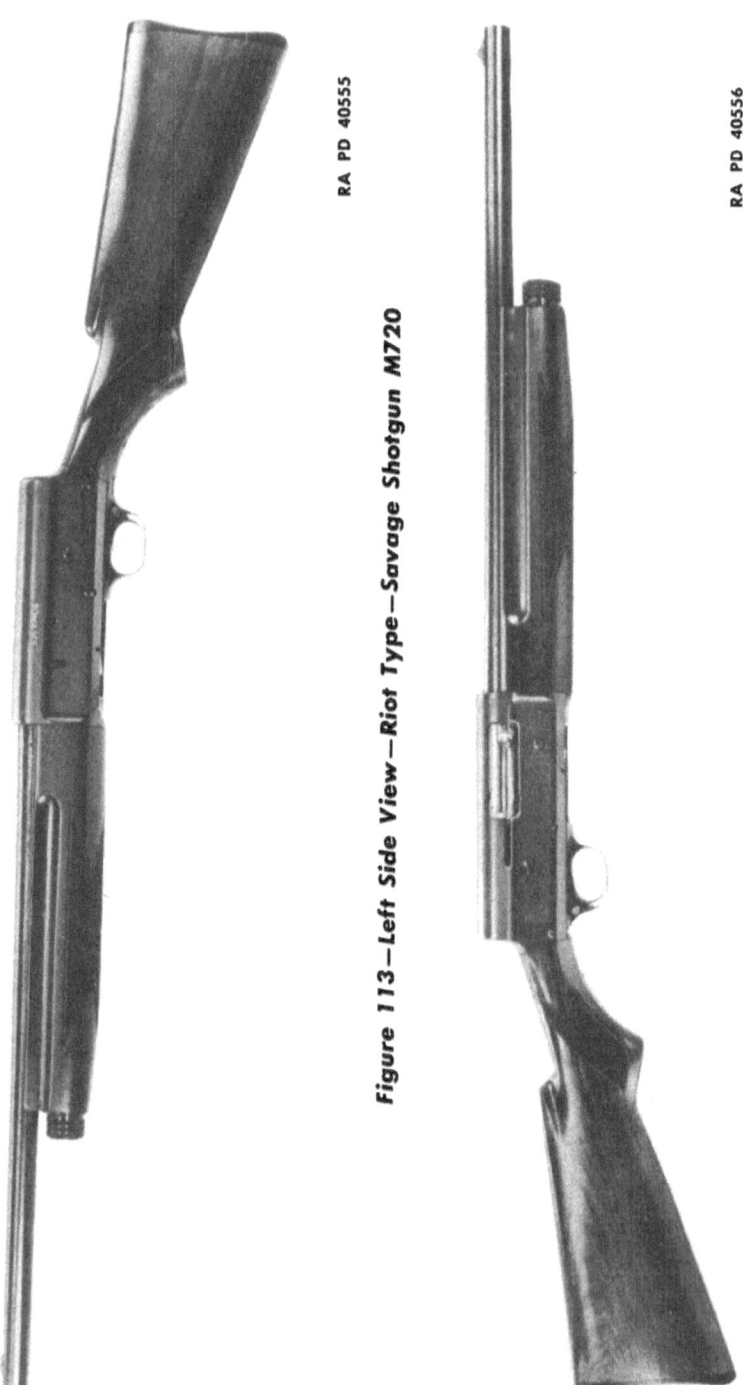

Figure 113 — Left Side View — Riot Type — Savage Shotgun M720

Figure 114 — Right Side View — Riot Type — Savage Shotgun M720

SAVAGE SHOTGUN, 12 GAGE, M720

d. This gun is furnished with barrels of different lengths and degrees of boring and may have other modifications of design. For convenience the gun will be classified as three types: riot, sporting skeet and sporting trap although variations of these types may occur.

(1) The riot-type gun (figs. 113 and 114) is usually furnished with a 20-inch plain barrel, bored full cylinder. Autoloading shotguns may not be equipped with bayonet attachments.

(2) The sporting skeet-type gun is usually furnished with a 26-inch plain or ribbed barrel, bored improved cylinder, and is without bayonet attachment or sling.

(3) The sporting trap-type gun is similar to the skeet-type gun but is furnished with a 30-inch barrel bored full choke, and is without bayonet attachment or sling.

e. **General Description** (figs. 115, 122, and 123). The general description of this gun is similar in every way to that of the Remington Gun M11 described in section VIII. The parts vary somewhat in shape in the two guns and they are not interchangeable. The position, removal, replacement, and functioning of the parts are generally the same as the Remington Gun M11. A few parts, however, are slightly different in design and nomenclature.

(1) The stock of the gun is fastened to the tang of the receiver by the tang screw passing through the tang of the trigger plate, through the stock grip, and threaded into the receiver tang. This screw is locked in place by a locking screw. To the rear end of the stock, a composition butt plate is fastened by two wood screws.

(2) The barrel has a semicylindrical extension threaded to the rear end which slides into the forward end of the receiver, and a ring guide brazed to the barrel tube which slides over the magazine tube. The magazine tube is screwed into the receiver at manufacture, and positioned by a stop screw. The recoil spring is mounted on the magazine tube together with a friction ring, friction piece, and spring. The positioning of these latter parts determines the amount of resistance (friction) set up against the force of recoil when the barrel is pulled to the rear by the breech bolt when locked to it. The barrel is reciprocated within the receiver and upon the magazine tube by the breech bolt which is locked to it during the rearward movement, and again at the end of the forward movement, by means of the locking block passing through an aperture in the barrel extension. A bead sight is driven into a ramp, pinned and brazed to the muzzle end of the barrel, and the ejector of the gun is riveted in a groove in the rear inner wall of the barrel extension.

TM 9-285

SHOTGUNS, ALL TYPES

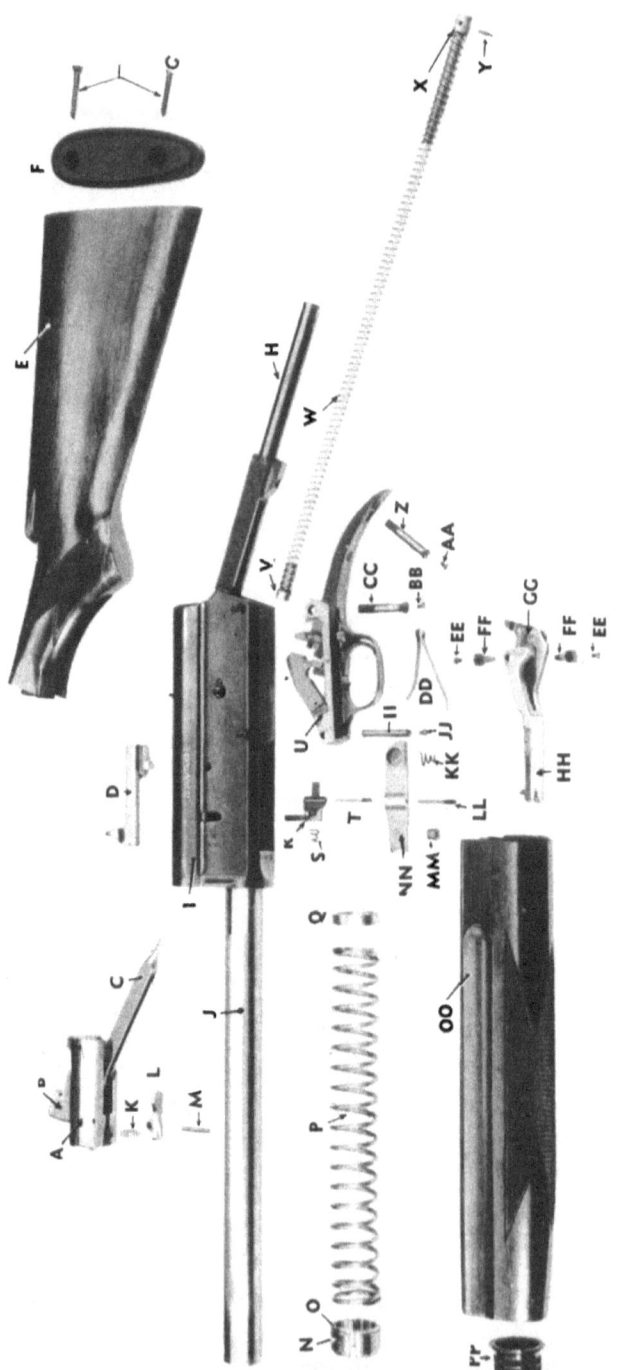

Figure 115—Disassembled View (Without Barrel)—Savage Shotgun M720

SAVAGE SHOTGUN, 12-GAGE, M720

A — BREECH BOLT
B — LOCKING BLOCK
C — LINK
D — OPERATING SLIDE
E — BUTT STOCK
F — BUTT PLATE
G — BUTT PLATE SCREW
H — ACTION SPRING TUBE
I — RECEIVER
J — MAGAZINE TUBE
K — LOCKING BLOCK LATCH SPRING
L — LOCKING BLOCK LATCH
M — LOCKING BLOCK LATCH PIN
N — FRICTION SPRING
O — FRICTION PIECE
P — RECOIL SPRING
Q — FRICTION RING
R — SHELL STOP
S — SHELL STOP SPRING
T — SHELL STOP SCREW
U — TRIGGER GUARD GROUP
V — ACTION SPRING FOLLOWER
W — ACTION SPRING
X — ACTION SPRING PLUG
Y — ACTION SPRING PLUG PIN
Z — BUTT STOCK SCREW
AA — BUTT STOCK SCREW LOCKING SCREW
BB — TRIGGER GUARD SCREW LOCKING SCREW
CC — TRIGGER GUARD SCREW
DD — CARRIER SPRING
EE — CARRIER SCREW LOCKING SCREW
FF — CARRIER SCREW
GG — CARRIER DOG
HH — CARRIER
II — TRIGGER GUARD PIN
JJ — TRIGGER GUARD PIN LOCKING SCREW
KK — CARRIER LATCH SPRING
LL — CARRIER LATCH SCREW
MM — CARRIER LATCH BUTTON
NN — CARRIER LATCH
OO — FORE-END
PP — MAGAZINE CAP

RA PD 40557A

Legend for Figure 115

SHOTGUNS, ALL TYPES

(3) The magazine tube is screwed into the forward end of the receiver and contains the magazine spring and follower retained by a split spring retainer, positioned in the forward end of the magazine.

(4) The fore end slides over the magazine tube and is held in position by the magazine cap and in turn limits the forward movement of the barrel. The magazine cap is locked in position by a spring plunger.

(5) The receiver contains the operating mechanism and to its lower rear end is fastened the trigger guard to which is mounted the firing mechanism. The trigger guard is secured by a screw, and a pin with locking screw. The receiver is open at the bottom for loading and at the right side for ejection of the fired shell cases. The rear of the receiver is formed into a tang in which is screwed the action spring tube. The tang and tube extend into the stock when the gun is assembled. The action spring tube houses the action spring which is stopped from rearward movement by a wooden stop pinned in the rear end of the tube. The action spring follower is positioned in the forward end of the action spring, and against it bears the rear end of the link when assembled. The action spring forces the breech bolt forward on the counterrecoil stroke by means of the link, which is pivoted to the locking block. The locking block slides radially in the bolt and is held in position by the firing pin which passes through bolt and locking block. It is this hinged construction of the link and locking block that enables the bolt to reciprocate horizontally while at an agle to the link and action spring. The vertical movement of the link operates the locking block latch positioned in the breech bolt, by which the locking block is blocked or freed to move radially in the bolt, when acted upon by the link.

(6) The carrier is pivoted in the rear of the receiver on two screws extending through the walls of the receiver and locked in place by locking screws. In the rear end of the carrier a dog is pivoted under spring tension supplied by a spring and plunger positioned behind it in the carrier. The carrier is raised by means of the dog which bears on the lower face of the breech bolt and lowered by action of the carrier spring positioned in the receiver.

(7) The carrier latch is pivoted in a groove in the inner right side of the receiver on a pin with a threaded head and is actuated by a coil spring positioned between it and the receiver. A button bearing on the latch extends through the right wall of the receiver, by which the latch can be manually disengaged from the carrier to free it.

(8) The shell stop is similarly pivoted and actuated in the left side of the receiver and acts to check the fed shell from entirely entering the receiver until the proper time.

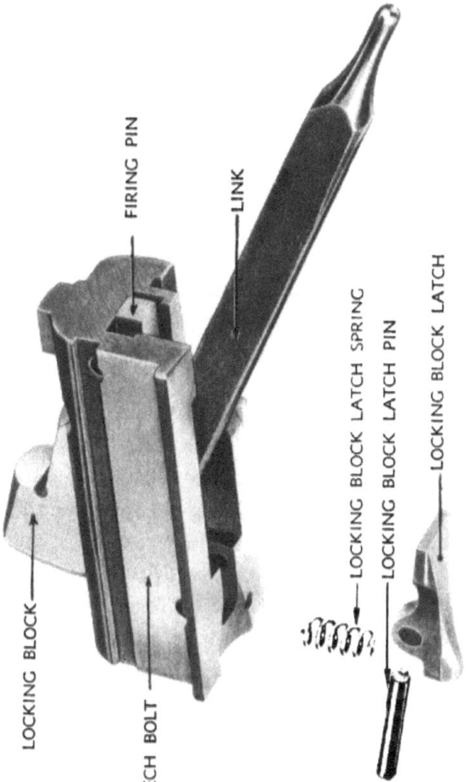

TM 9-285
61

SAVAGE SHOTGUN, 12-GAGE, M720

Figure 116 — Breech Bolt Group — Rear, Bottom and Left Side View — Locking Block Latch Disassembled — Savage Shotgun M720

SHOTGUNS, ALL TYPES

(9) The breech bolt (fig. 116) functions in guideways in the inner walls of the receiver and houses the firing pin which operates longitudinally through the length of the bolt and the locking block, and is cammed back into the bolt by the locking block as the block rotates downward. The extractor, positioned in the forward right face of the bolt, is of the claw type, and spring actuated. The locking block functions radially in guideways in the breech bolt and is positioned by the firing pin which passes through it and the breech bolt. To the lower rear end of the locking block, the link is pivoted by which the block is rotated upward to lock the breech bolt and barrel together when in the forward position. The firing pin is positioned by the firing pin stop pin passing laterally through the bolt and a groove in the firing pin. In the under side of the breech bolt, the locking block latch is pivoted and actuated by a coil spring, positioned between the latch and the bolt. The latch blocks the lower end of the locking block from forward (radial) movement until levered up by the link as the rear end of the link rises at the end of the forward movement. The operating slide, by which the breech bolt is manually retracted to load and cock the gun, is positioned in a groove in the right side of the bolt and reciprocates with it when the gun functions. The slide also functions with the link and carrier dog to unlock the breech bolt and barrel.

(10) The trigger guard group (fig. 117) is composed of the hammer, trigger, safety sear, mainspring, and trigger safety together with their components. The hammer is pivoted in the forward end of the trigger guard and is functioned by the mainspring screwed to the tang of the trigger guard and bearing on a roller in the lower end of the hammer. The mainspring is of the leaf type and is slotted to allow the upper end of the trigger which acts as the sear to protrude through it to engage the hammer. A lug extends from the upper rear face of the hammer in which there are two notches, one facing forward and the other to the rear. The trigger is pivoted in the rear end of the trigger guard, the lower end extending downward into the guard bow. The upper end of the trigger is U-shaped, the upper ends of the U forming hooks facing towards each other. When in the cocked position, the notched hammer lug lies between these hooks and is held by one or the other as explained hereafter. This feature prevents the gun from firing automatically as the trigger must be released and again pulled before the gun will fire. The safety sear is pivoted on a lug on the trigger guard directly over the trigger and acts as a block to the trigger during the functioning of the operating mechanism. The safety sear is spring actuated by a spring follower seated in the lug. The trigger safety is in the form of a cylinder, positioned to the rear of the trigger and operating laterally in the trigger guard to block or free the trigger for pulling to fire the gun.

SAVAGE SHOTGUN, 12-GAGE, M720

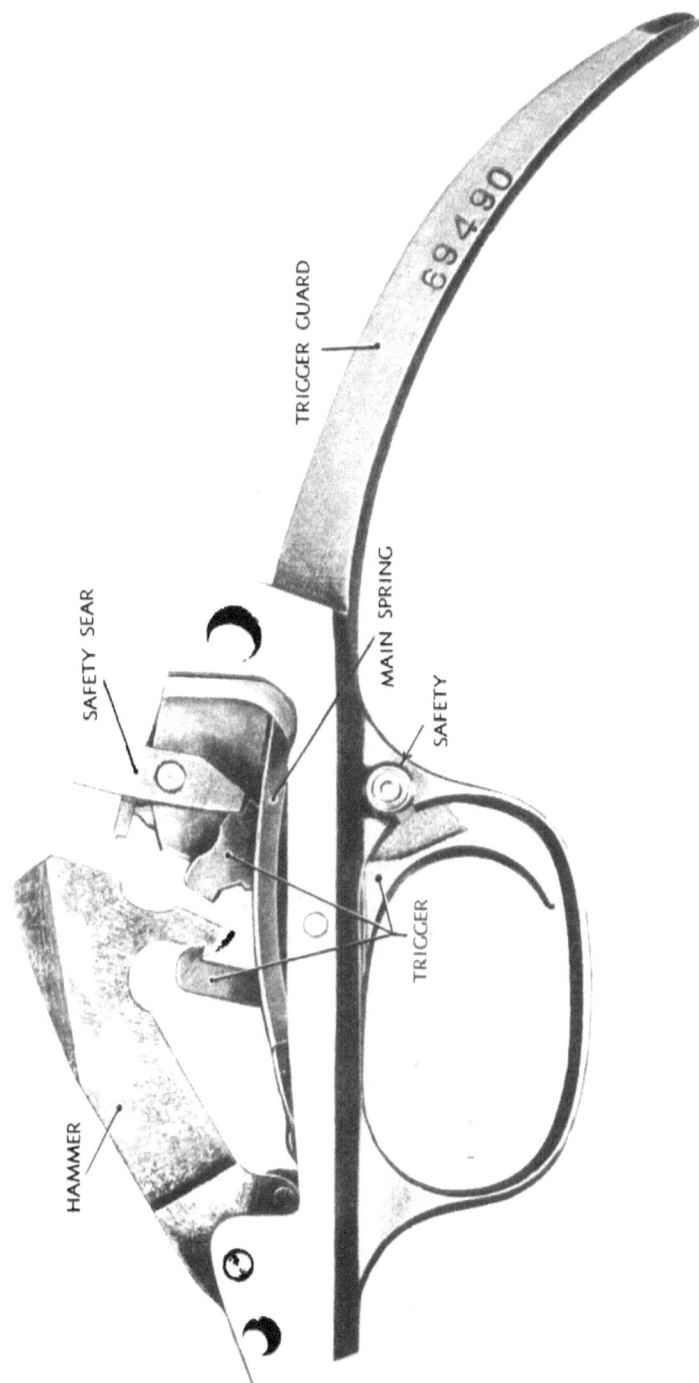

Figure 117 — Trigger Guard Group — Left Side View — Savage Shotgun M720

TM 9-285
62-63

SHOTGUNS, ALL TYPES

62. DATA.

Gage of bore	12
Boring of barrel—riot type	Cylinder
Boring of barrel—sporting skeet type	Improved cylinder
Boring of barrel—sporting trap type	Full choke
Type of action	Semiautomatic
Type of firing mechanism	Hammerless
Type of magazine	Tubular
Capacity of magazine	4 rounds
Length of barrel—riot type	20 in.
Length of barrel—sporting skeet type	26 in.
Length of barrel—sporting trap type	30 in.
Length of stock, receiver and fore-end assembled (approx.)	32½ in.
Length of assembled gun—riot type (approx.)	39½ in.
Length of assembled gun—sporting skeet type (approx.)	45½ in.
Length of assembled gun—sporting trap type (approx.)	49½ in.
Weight of assembled gun—riot type (approx.)	7⅝ lb
Weight of assembled gun—sporting skeet type (approx.)	8 lb
Weight of assembled gun—sporting trap type (approx.)	8¼ lb

63. OPERATION.

a. This gun is of the semiautomatic type as already explained and is loaded and cocked by the force of recoil after the first shot. The only action necessary, thereafter, to fire the gun is to pull the trigger. The first shell, however, must be loaded into the chamber and the hammer cocked by manual operation. This is accomplished by retraction of the breech bolt by means of the operating slide (fig. 120).

CAUTION: During operation, the muzzle of the gun should always be pointing at a safe spot and the finger kept outside of the trigger guard bow.

b. When the gun is operated as a repeater, the shells are loaded into the magazine, and a shell then transferred to the chamber, and the hammer cocked by manual retraction of the slide as already stated. When operating the gun as a single loader, the bolt can be retracted and hung in the rearward (open) position by retraction of the operating slide. This operation unlocks the bolt from the barrel, pulls it to the rear, and engages it with the dog of the carrier to hang it in the rearward position. The bolt is again released to move forward by pressing the carrier latch button, thus allowing the carrier to rise and the dog to release the bolt. The bolt springs forward and pushes the shell into the chamber of the barrel, and bolt and barrel are locked together as already explained.

c. The double hook on the trigger retains the hammer in the cocked position whether the trigger is released between shots or not. The trigger,

SAVAGE SHOTGUN, 12-GAGE, M720

however, must be released and again pulled before the hammer is released from the cocked position to fire the gun.

d. The safety sear blocks the trigger from being pulled during the rearward and forward movement of the operating mechanism. The safety sear is engaged with the trigger to block its being pulled by the spring action of the safety sear follower and disengaged from the trigger by the rear end of the link as the link locks the bolt and barrel together at the end of the forward movement. This feature prevents premature firing of the gun before the bolt and barrel are locked together in the forward position.

e. The trigger safety is positioned in the rear of the guard bow and operates laterally in the trigger guard (fig. 118). When pushed fully to the left, the trigger is free to be pulled and the gun fired. When pushed fully to the right, the trigger is blocked and cannot be pulled nor the gun fired. With a shell in the barrel chamber, it is best to block the trigger by pushing the safety to the right, as explained above, unless the gun is to be fired immediately.

f. **To Load the Magazine.**

(1) With breech bolt locked, press the carrier latch button fully inward to release the forward end of the carrier. Press shell nose down against carrier until it can be pushed forward against the magazine follower and into the rear of the magazine (fig. 119). As the base of the shell passes the end of carrier latch, pressure on latch button should be released slightly. Push the shell forward into the magazine until the base of the shell passes and is retained by the end of the carrier latch.

(2) Holding the carrier latch button depressed with the thumb, load another shell in the same manner, pressing it against the base of the first shell loaded, and push both shells forward into the magazine until the second shell is caught and held as above. When the carrier latch button is released, the rearmost shell will spring back and be held by the locking block latch which projects from the bottom of the bolt.

g. **To Unload the Magazine.** Set safety at safe by pushing it all the way to the right. Then, depressing carrier latch button, press carrier against breech bolt and holding carrier thus, pull operating handle part way to the rear to release the shell from the magazine onto carrier, and lift shell from receiver. The shell in the barrel (if loaded) will return to the barrel when the slide is released.

h. **To Load the Barrel from the Magazine.**

(1) With the magazine loaded, set safety at safe by pushing all the way to the right. Then, pull the operating slide fully to the rear and release. This action will cock the hammer and feed a shell from magazine

TM 9-285
SHOTGUNS, ALL TYPES

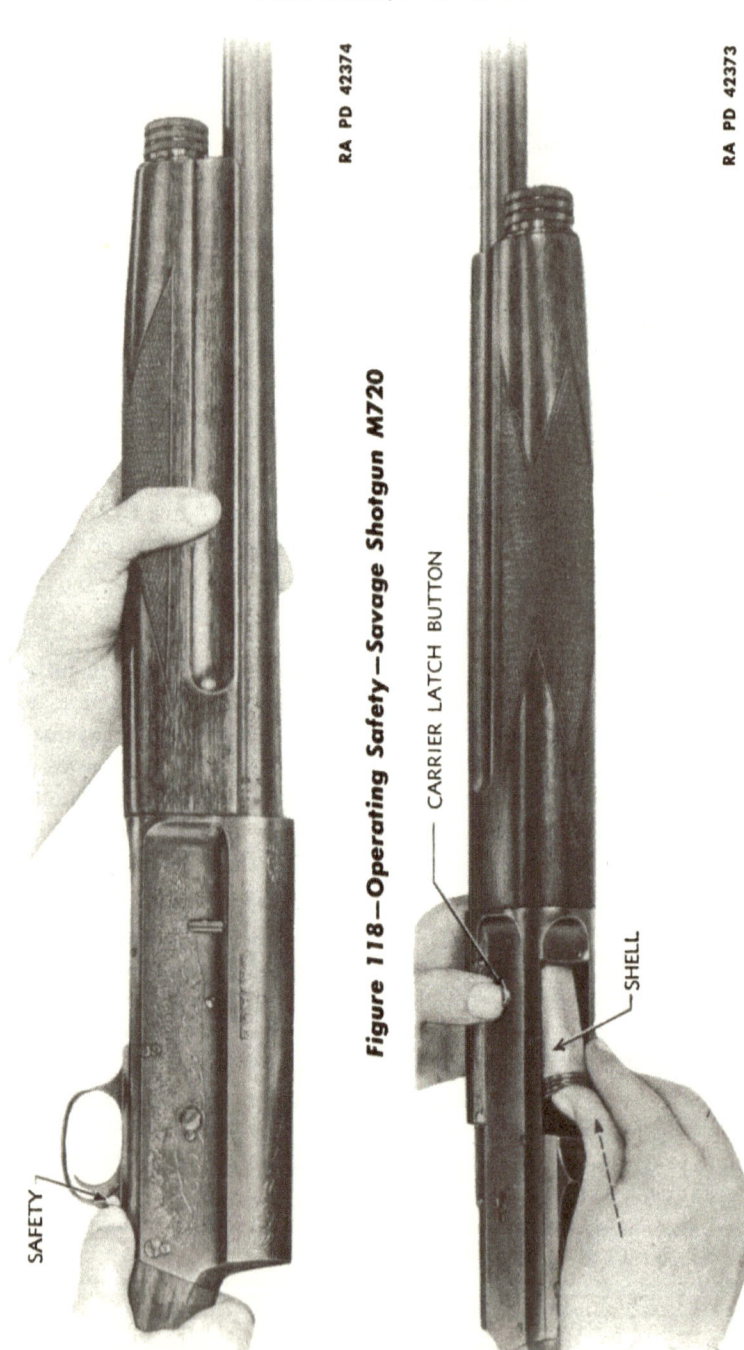

Figure 118 — Operating Safety — Savage Shotgun M720

Figure 119 — Loading Magazine While Depressing Carrier Latch Button — Savage Shotgun M720

SAVAGE SHOTGUN, 12-GAGE, M720

to receiver. When the breech bolt springs forward, it will push the shell into the barrel and close and lock the action.

CAUTION: The fingers should be kept clear of the ejection opening in the receiver as well as the path of the operating slide handle during this operation, to prevent injury to fingers due to the closing of the action.

(2) After a shell has been loaded from the magazine into the barrel, another shell may be loaded into the magazine (f above).

i. **To Load the Barrel Only.** With the magazine empty, set safety at safe by pushing all the way to the right and pull operating slide to rear and hang the bolt (fig. 120). Then, place a shell, nose forward, into ejection opening in the right side of the receiver and allow it to settle on the carrier. With fingers clear of ejection opening and path of the operating slide handle, press carrier latch button. The breech bolt will thus be released to spring forward; move the shell into the barrel and close and lock the gun. The barrel is now loaded and the hammer cocked. With the barrel loaded, it is best to allow the safety to remain at safe to block the trigger unless the gun is to be fired immediately.

j. **To Unload Barrel Without Unloading Magazine.** If it is necessary to unload the barrel without unloading the magazine, it can be accomplished as follows:

(1) Set safety at safe by pushing all the way to the right. Press carrier latch button and then push shell, which protrudes part way from the magazine, fully into the magazine until it is retained by the forward end of the carrier latch. Keep shell in magazine by continuing to press on latch button and retract operating slide to eject shell in barrel. Then, still pressing on carrier latch button, allow breech bolt to close slowly (fig. 121), watching to see that the bolt closes on an empty barrel chamber.

k. **To Completely Unload the Gun.** When gun is completely loaded (barrel and magazine) it should be unloaded as follows:

(1) Set safety at safe by pushing all the way to the right.

(2) Unload magazine first (g above).

(3) With magazine empty, unload the barrel chamber by pulling operating slide fully to rear, thus ejecting the shell in the barrel.

(4) Inspect barrel chamber and magazine to make sure gun is fully unloaded.

(5) When gun is fully unloaded, the action may be closed by holding operating handle, pressing carrier latch button, and easing breech bolt to the forward position.

TM 9-285

SHOTGUNS, ALL TYPES

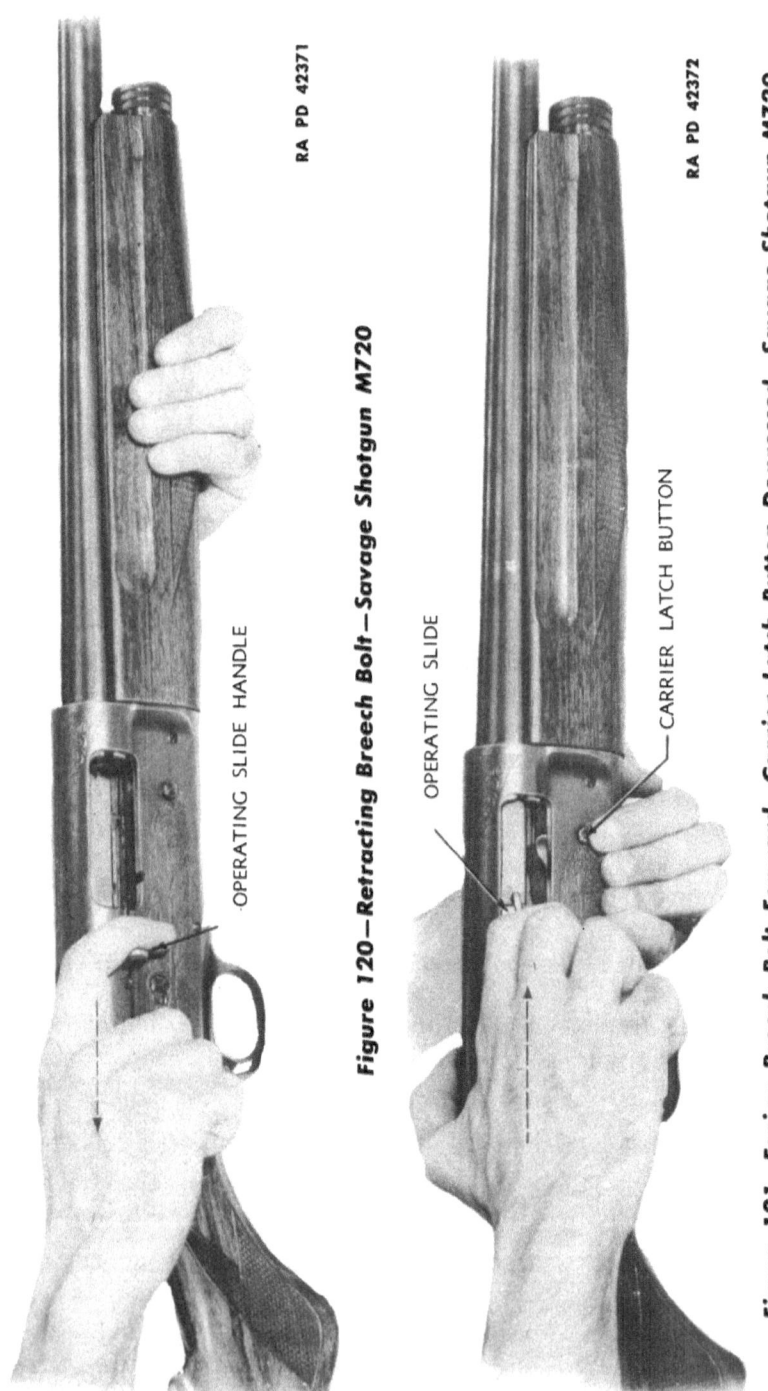

Figure 120 — Retracting Breech Bolt — Savage Shotgun M720

Figure 121 — Easing Breech Bolt Forward — Carrier Latch Button Depressed — Savage Shotgun M720

SAVAGE SHOTGUN, 12-GAGE, M720

64. FUNCTIONING.

a. As already explained, the functioning of this semiautomatic gun is accomplished through the force of recoil produced by the expanding powder gas acting through the shell against the breech bolt when the gun is fired. In this explanation of functioning, it is assumed that the gun is fully loaded and has just been fired.

b. As the powder gas expands, it forces the breech bolt to the rear along with the barrel which is locked to it by the locking block. As the bolt moves rearward, it pushes the link, which is pivoted to the rear end of the locking block, to the rear against the force of the expanded action spring with which it is engaged. The barrel is, at the same time, pulled to the rear by the breech bolt against the force of the expanded recoil spring and the resistance of the friction ring and piece, mounted on the magazine tube, acting against the rear face of the barrel guide attached to the barrel. Thus the action spring and recoil spring are compressed during the rearward movement by the bolt and barrel respectively. The resistance to compression of these springs and the friction set up by the contracted friction piece act together against the force of the expanding powder gas upon the breech bolt (par. 67).

c. The breech bolt and barrel move to the rear, locked together, to the end of the rearward movement. This point is roughly determined by the opposing force of the springs and friction piece as explained above which stops the groups and then starts them forward. As the breech bolt and barrel start forward, the dog on the rear of the carrier, which has been pressing upon the lower face of the bolt through the pressure of the carrier plunger spring, engages with the operating slide and checks the slide. The forward end of the carrier is prevented from rotating upward by the carrier latch. The operating slide, which has been bearing on the forward face of the link likewise checks the link. The bolt continuing forward, unlocks itself from the barrel extension as the locking block is rotated downward out of the locking aperture in the barrel by the momentarily held link pulling upon the locking block. As the bolt is unlocked the barrel springs forward, propelled by the action of the expanding recoil spring bearing on the rear face of the barrel guide. As the locking block rotates downward, it cams the firing pin back into the bolt, and holds it there until the bolt and barrel are again locked together.

d. The bolt is momentarily held and released as follows:

(1) When the bolt is unlocked, it is held to the rear by the engagement of the carrier dog with the operating slide and link as already explained. The bolt is held thus until the forward end of the carrier, which has been held in the lower position by the carrier latch, is freed to be rotated upward by the force of the action spring acting through the link and slide

SHOTGUNS, ALL TYPES

which are bearing upon the carrier dog. As the forward end of the carrier rises, the dog is disengaged from the slide and the bolt is freed to move forward.

(2) The carrier latch is disengaged from the carrier by the shell, which has been fed from the magazine into the receiver. This shell is held by the shell stop and released by the barrel as it springs forward when unlocked (f below). The released shell springs to the rear and strikes and depresses the rear end of the carrier latch, and thus disengages the latch from the carrier. The carrier is thus freed to be rotated upward as explained above.

(3) When the last shell has been fired, the bolt is held to the rear by the carrier dog as explained. There being no shell in the receiver to release the carrier by striking and depressing the carrier latch, the bolt will be held until the carrier is released, by manual pressure on the carrier latch button.

(4) The bolt thus released, as explained above, moves forward propelled by the action spring through the medium of the link and locking block. As the bolt is pushed forward, the link is at an angle to the axis of the bolt. In this position, the forward end of the link lies above the rear end of the locking block latch which is pivoted in the bolt and bearing on a shoulder on the locking block, preventing its rotation upward. The link thus pushes the bolt forward although the link is pivoted to the lower end of the locking block. As the bolt thus propelled nears the forward position, the rear end of the link rises until the link is nearly parallel to the axis of the bolt. As the rear end of the link rises, the forward end levers the locking block latch down from engagement with the locking block, and immediately begins to rotate the locking block upward in the bolt. This movement is so timed that the locking block is rotated fully through the bolt and into the locking aperture in the barrel extension as the bolt reaches the forward position, thus locking bolt and barrel together again. The rear lug on the locking block, striking the sloping rear end of the barrel extension, cushions the bolt and starts the locking block rotating upward.

e. When the bolt moves rearward on the recoil movement, the fired shell, held in position on the face of the bolt by the extractor on the right hand side, moves to the rear with the bolt and barrel. As the barrel is freed from the bolt at the beginning of the forward (counterrecoil) movement, it springs forward as explained. As the rear of the barrel extension passes the face of the bolt, the ejector pinned in the left wall of the barrel extension strikes the base of the fired shell held to the face of the bolt by the extractor and knocks it to the right out of the ejection opening in the receiver, pivoting it about the extractor.

SAVAGE SHOTGUN, 12-GAGE, M720

f. As the barrel extension continues forward, it strikes and disengages the shell stop which has been holding the next shell to be loaded. The shell springs rearward, propelled by the force of the magazine spring, to release the carrier latch and hence the bolt, and is then lifted to the bore line by the carrier which is rotated up by the link as explained in d above. The shell is pushed forward into the barrel chamber by the bolt, as it moves forward, and the carrier spring returns the carrier to the lower position.

g. As the rear end of the carrier latch is depressed by the shell and thus disengaged from the carrier, the forward end is rotated outward into the path of the next shell in the magazine. The latch holds the shell thus until the carrier moves downward again after having lifted the previous shell to the bore line. As the carrier moves downward, it releases the rear end of the carrier latch which it has been holding depressed. As the rear end springs outward, the forward end rotates in towards the receiver wall and thus frees the shell it has been holding. This shell springs to the rear and is held by the forward end of the locking block latch on the bolt, which has now reached the forward position, as the carrier reaches the lower position and thus retains the shell in the receiver. The above functioning prevents double feeding. The shell is held thus by the locking block latch until the bolt starts rearward again, when the shell follows it until caught and retained by the shell stop, from which it is subsequently released by the barrel extension on its forward movement as already explained.

h. The hammer is cocked by the link, which straddles it, on the rearward movement of the bolt. The sear hooks on the trigger (fig. 117) engage with notches in the hammer to hold it in the cocked position, whether the trigger has been released between shots or not. The safety sear blocks the trigger preventing its being pulled until disengaged by the link, as the bolt reaches the locked position. Detailed explanation of the functioning of the trigger is as follows:

(1) The upper part of the trigger is shaped like a U with two sear hooks facing each other. A lug projecting from the rear face of the hammer is notched on both sides to engage with the sear hooks on the trigger. The notched lug of the hammer rotates down between these hooks when cammed back by the link on the rearward movement of the bolt. If the trigger is in the forward (released) position, the forward hook engages with the forward notch in the hammer. If the trigger is in the rearward (pulled) position, the rear hook engages with the rear notch in the hammer. Thus the hammer is caught and held in the cocked position whether the trigger has been released or not. If the trigger has been held in the pulled position and the hammer caught by the rear hook, the trigger must be released and again pulled before the gun can be

SHOTGUNS, ALL TYPES

fired. Release of the trigger disengages the rear hook, but the hammer is immediately caught and held by the forward hook. When the trigger is again pulled, the hammer is released and rotates forward propelled by the force of the main spring, to strike the rear of the firing pin positioned in the breech bolt. With the locking block in the raised (locking) position, the firing pin is free to move forward in the bolt.

(2) The safety sear, however, has been engaged with the rear of the trigger by the action of its spring follower. In order to pull the trigger, the safety sear must be disengaged. This is accomplished by the rear end of the link, which straddles the safety sear, as it reaches the forward position to lock the bolt. This feature prevents the gun from being prematurely fired before the breech bolt is locked to the barrel.

i. The trigger safety functions laterally in the trigger guard just behind the trigger. A web on the rear face of the lower part of the trigger slips into a notch in the safety when the safety is pushed to the left and thus disengaged, allowing the trigger to be pulled. When the safety is shifted laterally to the right, the trigger web cannot enter the notch in the safety and therefore cannot be pulled.

65. REMOVAL OF GROUPS (figs. 115, 122, and 123).

a. Groups and parts should be removed and replaced in the order given below. Groups and parts when removed should be placed on a clean, flat surface and care observed to prevent loss of screws and small parts. Remove as follows:

(1) BARREL AND FORE END.

(a) With breech bolt in the forward position and gun held in vertical position, rest butt on solid surface.

(b) Push barrel back into receiver a short distance and hold in that position to relieve pressure of recoil spring on fore end; unscrew magazine cap at forward end of fore end (fig. 124).

(c) When magazine cap is fully disengaged from magazine, ease barrel upward out of receiver until free of pressure of recoil spring (fig. 125).

(d) Slide fore end up and off magazine and barrel and then pull barrel up and out of receiver.

(e) Note relative positions of friction piece, friction ring, and recoil spring as mounted on the magazine tube and if removed, replace in identical manner (par. 67) for assembly of these parts for varying recoil pressures.

(f) Replace fore end on magazine tube, press down against recoil spring, and screw on magazine cap to hold recoil spring, friction piece and friction ring in position. Exercise care not to lose friction piece spring.

CAUTION: When breech bolt is in rearward position, do not press

SAVAGE SHOTGUN, 12-GAGE, M720

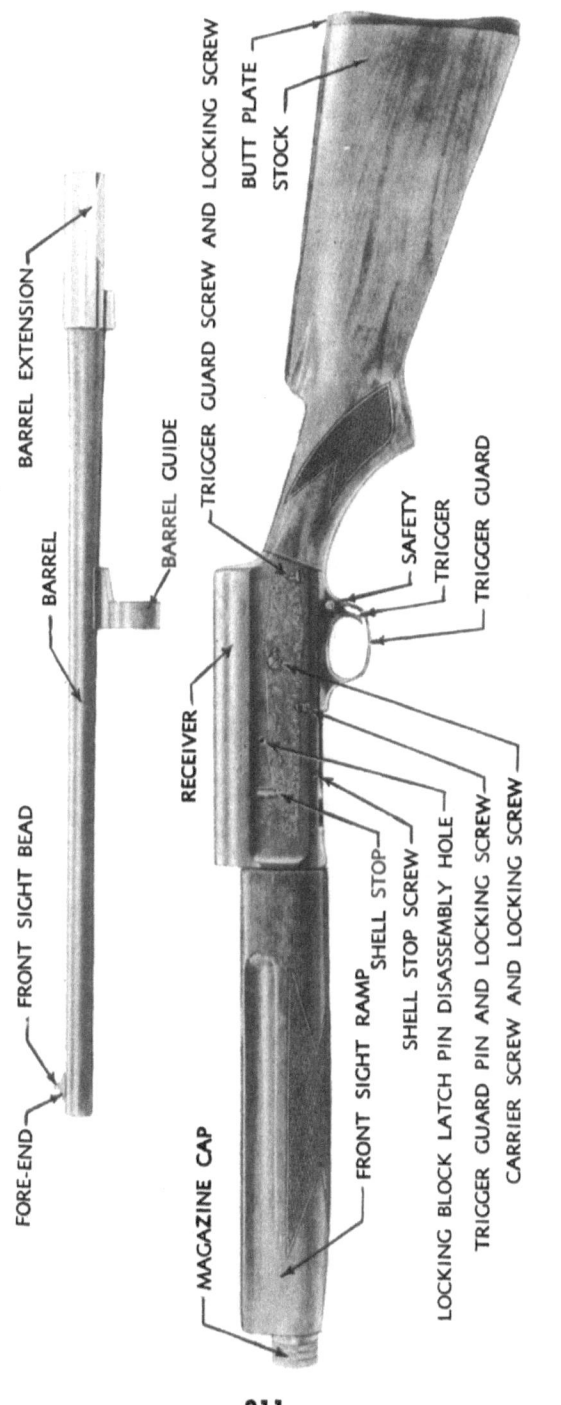

Figure 122—Gun Taken Down—Left Side View—Showing Location of Parts—Savage Shotgun M720

TM 9-285

SHOTGUNS, ALL TYPES

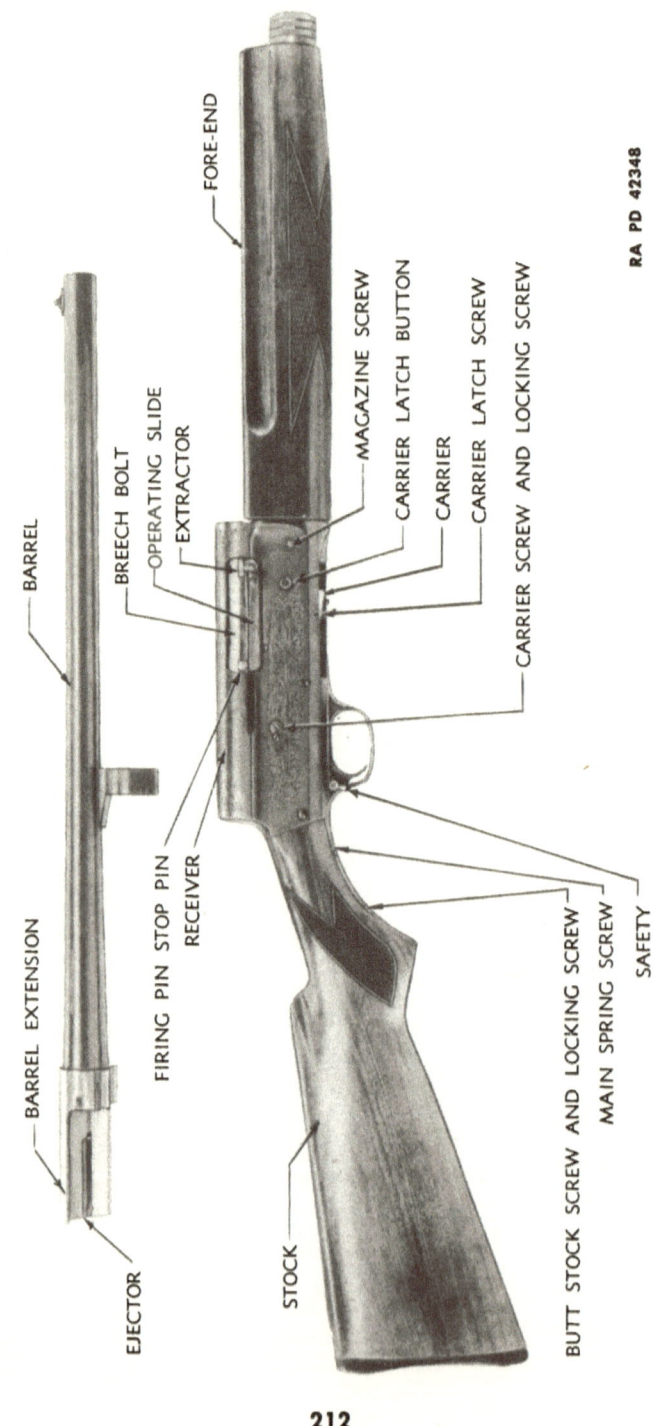

Figure 123 — Gun Taken Down — Right Side View — Showing Location of Parts — Savage Shotgun M720

TM 9-285
65

SAVAGE SHOTGUN, 12-GAGE, M720

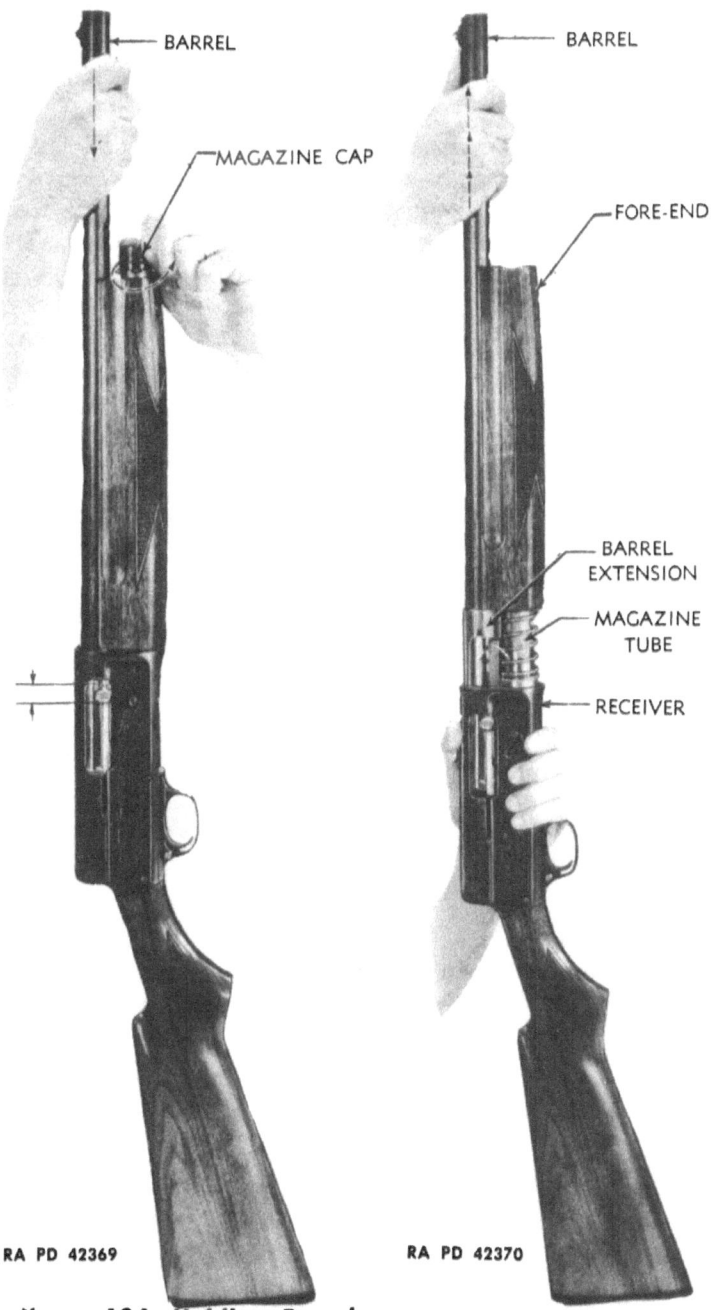

RA PD 42369

Figure 124—Holding Barrel—Slightly Retracted While Removing Magazine Cap—Savage Shotgun, M720

RA PD 42370

Figure 125—Easing Barrel (With Fore End) Out of Receiver—Savage Shotgun, M720

SHOTGUNS, ALL TYPES

carrier latch button allowing breech bolt to spring forward, with barrel removed from receiver. To remove tension from action spring, grasp operating slide handle firmly; then, press carrier latch button and ease bolt to forward position against force of action spring.

(2) BUTT STOCK. Due to expansion of action spring when breech bolt is removed it is necessary to remove this spring before removing the groups from the receiver. To remove the spring, the butt stock must first be removed. This is accomplished by removing the butt stock screw locking screw and then the butt stock screw from under side of trigger plate tang (rear screw in tang) and withdrawing butt stock to rear from receiver and action spring tube. (If withdrawal of butt stock is difficult, remove trigger guard pin, press carrier latch button to release carrier, and press forward end of trigger guard upward slightly into receiver.)

(3) TRIGGER GUARD GROUP.

(a) With stock removed, remove trigger guard pin locking screw from lower left hand side of receiver (near center) and, using pin drift, drive out trigger guard pin from right to left.

(b) Remove trigger guard screw from lower rear corner of left side of receiver, and pull trigger guard group downward out of receiver.

(4) CARRIER.

(a) Press forward end of carrier spring from under retaining stud and remove spring from pivot stud. Be careful spring does not fly out when disengaging forward end.

(b) Remove locking screws and then carrier screws from right and left sides of receiver and lift out carrier. Carrier screws are largest screws in sides of receiver, just to rear of center.

(5) ACTION SPRING.

(a) With butt stock, trigger plate, and carrier removed, press in the wooden plug at rear end of action spring tube and using pin drift, push out cross pin.

(b) Holding plug with finger against force of spring, withdraw drift slowly and ease plug and spring to rear out of tube. When fully expanded pull plug, spring, and spring follower to rear out of tube.

(6) BREECH BOLT AND LINK. Recoil spring, friction piece, and ring must be moved forward on the magazine tube before breech bolt can be removed from receiver.

(a) Move breech bolt until locking block latch pin alines with hole in left side of receiver to rear of shell stop. Then, using straight pin drift, push clear through breech bolt and receiver; then, withdraw pin drift slowly, at same time holding down locking block latch against

SAVAGE SHOTGUN, 12-GAGE, M720

force of latch spring. When drift is removed, ease latch upward and remove latch and latch spring.

(b) Swing rear end of link downward to clear operating slide, and move breech bolt forward out of receiver with the link attached. The operating slide will slide out of the breech bolt as it moves forward and the handle of the slide strikes the forward edge of the ejection opening in the receiver.

66. REPLACEMENT OF GROUPS.

a. Groups and parts should be thoroughly cleaned, lightly oiled, and lubricated, if necessary, before replacing. Replace as follows:

(1) BREECH BOLT AND LINK.

(a) With recoil spring, friction piece, and ring moved forward on the magazine tube, slide the link and breech bolt, link first, into the forward end of the receiver, so that guides on lower face of breech bolt mate with guideways in receiver.

(b) Move breech bolt to rear into receiver until half the bolt has entered. Then, holding operating slide by the handle, slip rear end into receiver, to rear of bolt, through ejection opening in right side. Hold slide parallel to bolt and move bolt to rear, at the same time mating guides on slide with guideways in right side of bolt. As slide enters bolt, move bolt to rear thus seating slide in bolt. The link must be pointed downward to accomplish the mating.

(c) Slide recoil spring, friction piece, and friction ring to rear on magazine tube, in their proper order and position, and replace fore end to hold in place (par. 65 a (1) (f)).

(d) Slide breech bolt to rear until locking block latch pin holes in bolt and receiver aline, insert latch pin through left side of receiver and left wall of bolt to hold in position; then, replace latch spring in seat in bolt and place latch on top of spring, with long flat end to rear and drilled lug down. Press latch down on spring; aline hole in latch with holes in receiver and bolt and push pin through latch and bolt until flush with both sides of bolt. Move bolt to see that pin does not interfere with receiver.

(2) CARRIER.

(a) With receiver bottom side up, move breech bolt forward and push link towards top of receiver.

(b) With carrier dog up and to rear, slide carrier beneath spring retaining stud, into receiver, and aline screw holes in carrier with those in receiver. Replace carrier screws; then, turn screws until cuts in screw heads aline with countersinks for locking screws and replace locking

SHOTGUNS, ALL TYPES

screws. Catch threads in both carrier screws before tightening either screw. Be careful not to cross threads.

(c) Slide carrier spring onto the pivot stud in left rear of receiver so that short leaf rests on rear of carrier. Then, press down long leaf of spring and slip under stud head just forward of pivot stud. Hold spring to prevent slipping from pivot stud while positioning long leaf.

(3) ACTION SPRING.

(a) With breech bolt forward, insert action spring, follower first (assembled to spring), into rear end of action spring tube. Push through tube and mate rear nose of link with indent in follower.

(b) Press action spring into tube until head of wooden plug (assembled to spring) is flush with end of tube. Turn plug until pin hole in plug alines with that in tube and insert plug pin. Still holding plug, push pin through until flush with tube. A pin drift may be used to hold plug while inserting pin.

(4) TRIGGER GUARD GROUP.

(a) With the breech bolt forward and the hammer cocked, press trigger guard group with tang to rear horizontally upward into the rear under side of the receiver and adjust until pin and screw holes in trigger plate and receiver aline. If guard does not seat easily, retract operating slide slightly to allow safety sear to enter slot in link.

(b) Insert trigger guard pin from left side and push through until head is flush with face of receiver.

(c) Using screwdriver in slotted head of pin, turn pin until countersink in pin head alines with locking screw countersink in receiver. Then, replace locking screw.

(d) Insert trigger guard screw from left side of receiver; catch threads and turn down tightly. Aline cut in screw head with locking screw countersink in receiver and replace locking screw.

(5) BUTT STOCK.

(a) Push butt stock on over action spring tube until stock fits snugly and evenly against rear face of receiver and butt stock screw hole in trigger guard tang and stock aline. Stock may be seated by striking butt smartly with heel of hand. Do not strike butt on hard surface.

(b) Replace butt stock screw and screw in tightly until cut in screw head alines with locking screw countersink in tang; then replace locking screw. If replacement of butt stock is difficult, remove trigger guard pin. Replace pin after assembly.

(6) BARREL AND FORE END.

(a) With breech bolt forward, fore end removed from magazine tube,

SAVAGE SHOTGUN, 12-GAGE, M720

and recoil spring, friction piece, and friction ring in their proper order and position on magazine tube, hold gun in vertical position and rest butt of stock on firm surface.

(b) Slide the barrel extension into open end of receiver so that guides on barrel extension mate with guideways in receiver and barrel guide slips over magazine tube and bears on friction assembly on tube.

(c) Slide fore end on barrel and magazine tube as far as it will go.

(d) Grasp muzzle end of barrel and push barrel down into receiver against spring force, until the barrel extension is wholly within the receiver. Then, press fore end down until it mates evenly with receiver and engage and screw down magazine cap on forward end of magazine tube until tight. Be sure fore end is fully mated flush with receiver and magazine cap is tight.

(e) Release barrel and operate gun to test assembly.

67. USE OF FRICTION RING.

a. Proper use of the friction ring and piece will reduce recoil and prevent excessive wear of parts. The adjustment of these parts in this (Savage) gun varies slightly from that of the Remington guns. The parts also vary slightly in design; the friction ring on the Remington guns has an inside bevel on one face and an outside bevel on the other, while that of this (Savage) gun has an inside bevel on one face only.

b. To change friction adjustment, remove fore end and barrel and place the recoil spring, friction piece (with friction spring assembled), and friction ring as prescribed below.

(1) FOR HEAVY LOADS. When heavy loads are used ranging from 3¼ to 3¾ drams (or equivalent) of powder, assemble as follows:

(a) Place the recoil spring on the magazine tube first, so that it bears directly against the receiver.

(b) Place the friction ring next on the magazine tube with the inside bevel facing forward.

(c) Place the friction piece (with friction spring assembled) on the magazine tube ahead of the friction ring.

NOTE: Refer to figure 126, position 1.

(2) FOR LIGHT LOADS. When light loads are used, 3 drams (or equivalent) or less of powder, assemble as follows:

(a) Place friction ring on the magazine tube first, with inside bevel facing to rear towards receiver.

(b) Place the recoil spring on the magazine tube next.

(c) Place the friction piece (with friction spring assembled) ahead of recoil spring.

NOTE: Refer to figure 126, position 2.

TM 9-285

SHOTGUNS, ALL TYPES

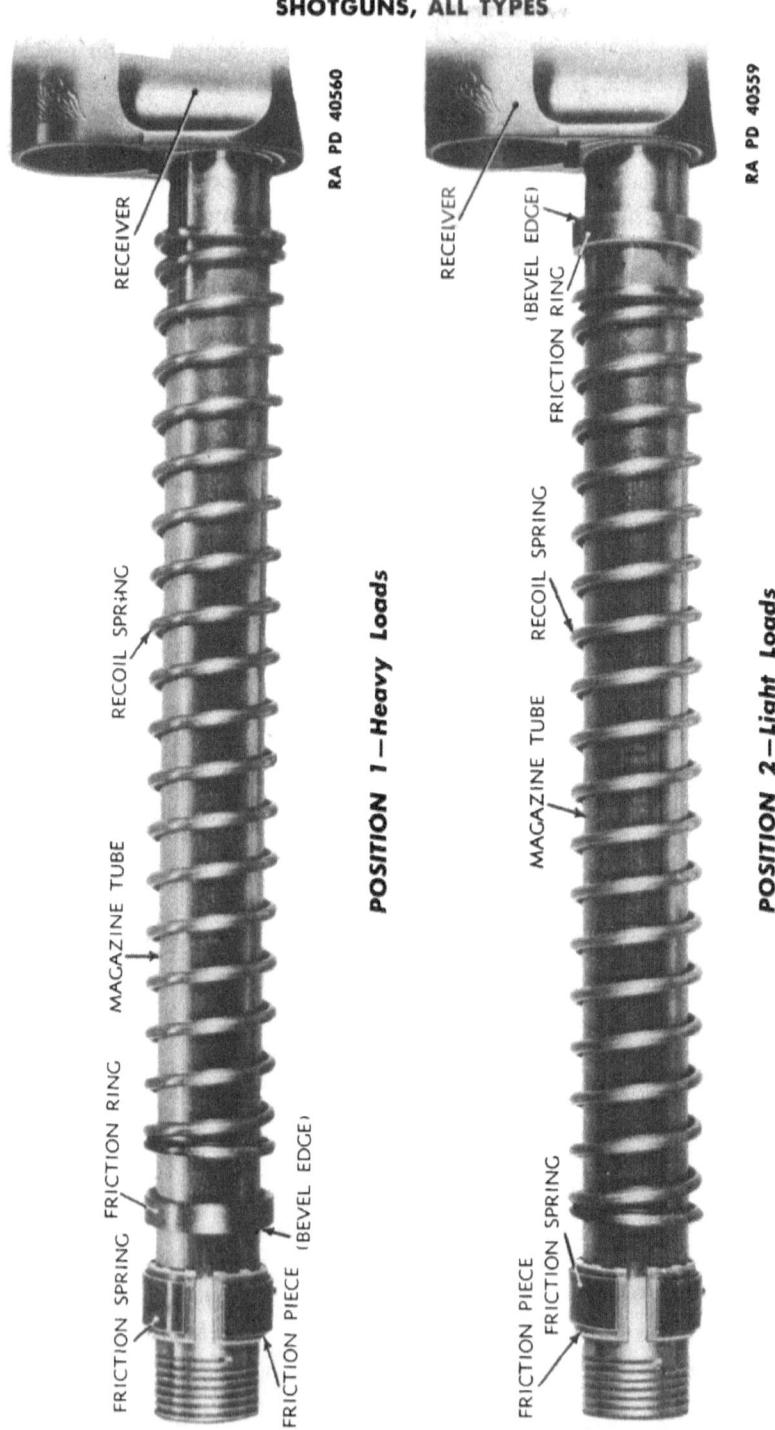

Figure 126—Recoil Adjustments—Savage Shotgun M720—

SAVAGE SHOTGUN, 12-GAGE, M720

c. Magazine tube should be kept free from foreign matter and rust, and lightly oiled. If gun does not operate smoothly and freely, lubricate friction piece and ring lightly. If friction is too great, gun may fail to eject shell when fired. If gun fails to eject, clean and lubricate magazine tube and friction piece and ring. If this does not correct fault and friction is adjusted for heavy loads, change adjustment to that for light loads, and test. If action slams too much with this lighter friction adjustment, turn gun over to ordnance personnel for correction.

68. FIELD INSPECTION.

a. With the gun completely assembled, test the mechanism for proper functioning. Fired shells may often be used for testing when dummy shells are not available, by turning in the uncrimped end so that the length of the shell will approximate that of a live shell. Use of live shells for testing is prohibited.

CAUTION: Keep fingers clear of ejection opening and path of operating slide handle when operating gun, as bolt moves forward with considerable force when released and may cause injury to operator. Be sure gun is fully unloaded before inspection.

b. **Operate the Gun as Follows:**

(1) Grasping fore end with left hand and operating slide handle with right hand, pull breech bolt all the way to the rear until bolt is caught and hung by carrier dog. Barrel should remain in forward position.

(2) With bolt hung in rearward position, press carrier latch release button and allow the bolt to spring forward. The bolt should spring forward smartly to the extreme forward position to lock with the barrel.

(3) Load a dummy shell into the magazine (par. 63 f). Retract breech bolt fully by means of operating slide and attempt to hang bolt in rearward position. Bolt should not hang, due to shell bearing on carrier latch as explained (par. 64 d).

(4) Allow bolt to spring forward to load the shell into the barrel chamber. Retract bolt fully by means of operating slide to eject the shell and attempt to hang bolt. Bolt should hang, as there is no longer a shell in the receiver to bear upon the carrier latch.

(5) Test gun for doubling as follows:

(a) With gun fully unloaded, pull bolt fully to the rear by means of the operating slide handle, until it is hung by the carrier dog.

(b) Hold bolt thus with right hand and with the left hand press carrier latch button to release the bolt and ease bolt forward just enough to release it from the hung position (about ¼ inch). Then, with left hand, pull the trigger.

SHOTGUNS, ALL TYPES

(c) Holding trigger retracted ease bolt slowly forward to the locked position.

(d) Then, ease pull on trigger slowly until entirely removed and the trigger moves forward. The hammer should not release to fire the gun during the slow release of the trigger.

(e) When trigger is fully released, pull it again. The hammer should release to fire the gun when the trigger is thus pulled. This test is to insure that the gun will not double (fire automatically) if the trigger is not released. To insure the gun will not double, it is desirable that the movement of the trigger at the center point of finger contact be not less than $\frac{1}{16}$ inch (par. 68 c (10)).

(6) Retract the bolt fully and, then, holding operating slide, press carrier latch release button and allow the bolt to close slowly. Release and attempt to pull the trigger at intervals during the closing of the bolt. The trigger should not pull to fire the gun until the operating slide handle is within $\frac{1}{16}$ to $\frac{1}{8}$ inch from the normal forward position. This test is to make sure that the gun will not fire until the locking block is engaged with the recoil shoulder of the barrel extension and bolt and barrel thus locked together.

(7) Retract bolt fully and then allow it to move fully forward to the locked position, thereby cocking the hammer. Then push the trigger safety all the way to the right and attempt to pull the trigger. The trigger should not pull, nor the hammer release to fire the gun.

(8) Push the safety all the way to the left, and attempt to pull the trigger. The trigger should pull and the hammer release to fire the gun.

c. When the gun does not operate and function smoothly when tested as above, damaged or improperly assembled parts are indicated as follows:

(1) BOLT DOES NOT HANG IN REARWARD POSITION WITH GUN UNLOADED. May be due to broken or worn carrier dog, worn retaining notch of operating slide or link shoulder, broken or missing carrier latch spring, broken carrier latch or foreign matter under forward end of latch.

(2) BOLT DOES NOT SPRING FORWARD SMARTLY WHEN RELEASED. May be due to broken action spring, link broken or improperly seated in action spring follower, jammed or broken carrier or carrier latch, broken locking block latch, or broken or missing latch spring.

(3) BOLT DOES NOT LOCK TO BARREL. May be due to foreign matter in receiver or barrel extension, burred locking block or locking apertures in barrel extension, or broken or jammed locking block latch spring.

SAVAGE SHOTGUN, 12-GAGE, M720

(4) CARRIER DOES NOT FUNCTION TO HANG BOLT OR LIFT SHELL. May be due to broken carrier dog or follower spring, jammed or broken carrier latch, foreign matter in mechanism, or worn or broken operating slide.

(5) CARRIER DOES NOT RETURN TO LOWER POSITION. May be due to broken or improperly assembled carrier spring, or jammed or broken carrier latch.

(6) SHELL DOES NOT SPRING FROM MAGAZINE WHEN RELEASED. May be due to broken magazine spring, dented tube, bent follower, or foreign matter or rust in tube.

(7) HAMMER DOES NOT COCK, OR SLIPS. May be due to worn or broken hooks on trigger, or notches in hammer, broken trigger or mainspring, or foreign matter in mechanism.

(8) SAFETY SEAR DOES NOT RELEASE TRIGGER WHEN BOLT IS LOCKED. May be due to broken or jammed safety sear spring follower, or damaged safety sear.

(9) TRIGGER SAFETY DOES NOT OPERATE. May be due to burred web on rear of trigger, burs in slot in safety, broken or bent leaf on trigger spring, which bears on safety ball, or foreign matter in aperture.

(10) HAMMER IS RELEASED WHEN TESTED (h (5) above). May be due to worn or broken hooks on hammer or trigger, foreign matter in notches or spread or broken upper U-end of trigger.

(11) THE FOLLOWING MALFUNCTIONS MAY OCCUR WHEN GUN IS FIRED:

(a) Barrel Is Not Unlocked from Breech Bolt. May be due to broken or damaged carrier dog or carrier latch, worn operating slide notch, or broken slide.

(b) Shells Do Not Eject. May be due to broken ejector in barrel extension, broken extractor or incorrect friction adjustment (par. 67 c).

(c) Shell Stop Does Not Function. May be due to broken stop or broken or missing stop spring.

(d) Two Shells Enter Receiver at Once (Double Feeding). May be due to broken carrier latch or carrier dog.

d. Inspect barrel and test trigger pull (par. 3 n).

e. In addition to inspection of the gun for operation and functioning, the gun should be inspected generally for condition, and defects noted. Attention should be directed to such defects as cracked wooden parts, cracked or deformed metal parts, dented or rusted magazine tube, loose screws and pins, loose or binding parts or assemblies, loose barrel or magazine, loose stock or butt plate, missing locking screws, rust, dents, burs or excessive wear of parts. If defects are such that early malfunc-

SHOTGUNS, ALL TYPES

tion of the gun is indicated, the gun should be turned over to ordnance personnel for inspection and correction.

f. Where defects and malfunctions cannot be remedied by cleaning, lubrication, and simple adjustments of assembly, which lie within the scope of using troops, the gun should be turned over to ordnance personnel for a thorough inspection, correction, and/or repair.

g. Removal of burs on working parts, trigger adjustments, and like corrections should not be attempted by using troops, as stoning of parts must be exacting, the angle of the faces concerned must not be changed, and volume of metal must not be materially reduced.

h. In addition to inspection of the barrel (d above), it should be inspected for looseness and alinement in the barrel extension. The draw marks on barrel and barrel extension should be in alinement, as misalinement of these parts will cause binding of the barrel guide with the magazine tube and affect the functioning of the gun when assembled. If misalinement is evident, the gun should be turned over to ordnance personnel for correction.

i. If shell appears unnecessarily loose in chamber with breech bolt locked, the gun should be turned over to ordnance personnel to be checked for headspace and worn locking block.

j. Adjustment and maintenance of the gun, in the case of using troops, is limited to the removal and replacement of the parts and groups of parts (pars. 65 and 66), together with cleaning, lubrication, and such adjustments as are necessary in assembling the gun as outlined.

69. CLEANING AND LUBRICATION.

a. Cleaning, oiling, and lubrication may be accomplished in a manner similar to that described for the Winchester Gun M97 (par. 11). Attention should be given to corresponding parts and surfaces when lubricating.

b. The fore end and barrel should be removed from the receiver for cleaning the bore, and for thorough cleaning, the groups should be removed from the receiver and assemblies cleaned, oiled, and lubricated as directed.

c. Points to lubricate are:

(1) Outer surface of barrel extension and barrel extension guides.

(2) Friction piece and friction ring, when necessary, where they bear on magazine tube. Tube should be kept clean and lightly oiled. Too much oil will reduce friction and prevent proper functioning of gun (par. 67 c).

(3) Breech bolt guides and surface, occasionally.

(4) Link pin, connecting locking block with link.

SAVAGE SHOTGUN, 12-GAGE, M720

(5) Locking block guides.

(6) Trigger and hammer pins.

(7) Carrier (trunnion) screws.

(8) Action spring follower. Spring should be removed occasionally and tube cleaned and oiled.

(9) Safety sear stud and spring follower.

d. Oiling and lubrication should be light, as excess oil collects foreign matter and powder residue which will become gummy, impede functioning of gun, and produce undue wear of parts. In very cold climates, oiling and lubrication should be reduced to a minimum. Only surfaces showing signs of wear should be lightly oiled. Refer to "Special Maintenance," section XI.

TM 9-285
70-71

SHOTGUNS, ALL TYPES

Section X

STOPPAGES AND IMMEDIATE ACTION

	Paragraph
General	70
Stoppages and immediate action	71
Malfunctions, possible cause and correction	72
To remove a shell from the chamber	73
To remove a shell from the magazine	74

70. GENERAL.

   a. A malfunction is the failure of a part or assembly of the gun to perform the operation it is designed for. A malfunction will result in a stoppage, or failure of the gun mechanism to perform its function when operated. Thus, the failure to extract a fired shell may be termed a malfunction and the consequent blocking of the barrel chamber preventing the loading of a live shell may be termed a stoppage. Such a stoppage may be due to a malfunction of the mechanism or faulty ammunition such as rupture of a shell.

   b. Immediate action is the term applied to the automatic application of a predetermined method of relieving the stoppage and thus putting the gun in action as soon as possible.

   c. Stoppages due to improper operation of the gun or faulty ammunition may, as a rule, be quickly relieved through immediate action. Stoppages caused by malfunction of parts due to wear, breakage, or improper assembly, cannot as a rule be immediately relieved and to such immediate action does not apply. Both types of stoppages will be covered herein. The failure, its probable cause, and suggested correction will be tabulated in order. Other stoppages may occur and should be dealt with by analyzing their cause and, if possible, correcting it. A thorough understanding of the proper assembly, operation, and mechanical functioning of the gun is essential in determining the cause of malfunctions and stoppages.

71. STOPPAGES AND IMMEDIATE ACTION.

   a. Stoppages and immediate action, as treated herein, comprises only such procedure as can be immediately applied through direct operation to put a gun in action which has become jammed or has failed to function. If immediate action does not put the gun in action, it should be inspected and remedied (par. 72).

   b. In the slide action type of shotgun covered herein, proper operation is of the greatest importance to the functioning of the gun. Reciprocation

224

STOPPAGES AND IMMEDIATE ACTION

of the slide handle should be complete and positive. The slide handle should be retracted smartly and fully to insure extraction and ejection of the fired shell and positive locking and/or cocking of the hammer. The slide handle should be pushed forward smartly and fully to insure positive chambering of the shell, locking of the breech bolt, and engagement of the slide lock. Unnecessary slamming of the mechanism should be avoided. In the autoloading guns the operating slide should be pulled fully to the rear and then released on the initial loading for the same general reasons.

c. The table of Stoppages and Immediate Action given below is intended for guidance only and should be applied to the gun in question as applicable. The terms slide handle, slide lock, and safety are applicable to like parts on the slide action guns covered herein. The term operating slide is applicable to the autoloading guns covered herein. Knowledge of the proper operation of the gun is assumed.

d. **Stoppages and Application of Immediate Action.**

Stoppages	Immediate Action
Gun fails to fire.	Reciprocate action slide handle or retract and release operating slide and fire.
Gun still fails to fire.	Glance into magazine. If empty, reload gun and fire.
Fed shell fails to chamber.	Retract action slide handle, or operating slide, remove fed shell from receiver; check chamber for shell or ruptured case. Remove obstruction; load and fire.
Gun fails to fire (shell fed and ejected when gun is operated).	Inspect primers of ejected shells. If normally indented, faulty ammunition is indicated. Reload gun and fire. If lightly indented, broken, or obstructed, firing pin is indicated.
Gun fails to fire (trigger fails to pull).	Check position of safety. Shift to fire position if on safe. Release and again pull trigger in Remington Gun M31 and autoloading guns.
Gun fails to fire (hammer unreleased).	Check position of safety. Reciprocate action slide handle or retract and release operating slide, and fire.

SHOTGUNS, ALL TYPES

CAUTION: When gun fails to fire, point muzzle to safe spot and turn ejection opening towards ground before operating to minimize effect of a possible hangfire shell (par. 93).

72. MALFUNCTIONS, POSSIBLE CAUSE AND CORRECTION.

a. Malfunctions may be caused by broken, damaged, or incorrectly assembled parts, faulty ammunition, incorrect operation of the gun, or the presence of fouling or foreign matter in the mechanism. Guns should be kept clean and lubricated at all times and inspected for operation and functioning at regular intervals or after encountering extreme conditions such as dust storms or the like. The receivers of slide action and autoloading shotguns are comparatively open and foreign matter can easily enter and clog the mechanisms. A clean, properly lubricated gun, inspected at frequent intervals, and ammunition stored and handled to prevent deterioration, is the best insurance against malfunctions and stoppages (pars. 92 and 94).

b. As organization spare parts are not provided for these guns, only such adjustments and minor repairs may be performed as lies within the scope of using troops and can be accomplished without replacement of parts. A gun which fails to operate or function properly should be turned over to an officer or qualified mechanic for inspection. If such inspection discloses malfunction due to damaged or broken parts, the gun should be turned over to ordnance personnel for repair.

CAUTION: When handling loaded guns for malfunction, the safety should be immediately placed at safe and gun handled with extreme care until fully unloaded.

c. The slide action guns covered in this technical manual are in general similar in basic design, operation, and functioning. The autoloading guns covered are practically identical in these respects. As already explained, nomenclature of similar parts varies somewhat with the make of gun. Malfunctions will therefore be covered generally herein as a guide only to be applied to the specific gun if and when applicable.

d. The possible malfunctions listed herein and their probable causes and remedies are not subject to immediate action and are within the scope of using troops only as indicated. Before inspecting guns, slide safety to safe position and fully unload gun.

e. **Failure to Fire.** Failure to fire may be due to a number of causes, as follows:

(1) EMPTY CHAMBER.

(a) Due to failure of operator to load, failure of operating mechanism to feed, empty or jammed magazine.

STOPPAGES AND IMMEDIATE ACTION

(b) To Remedy. Pump shell into chamber; load magazine; inspect and clear magazine (par. 74).

(2) Failure of Shell.

(a) Due to faulty primer, short or broken firing pin or foreign matter in firing pin aperture, broken hammer, broken or weak hammer spring.

(b) To Remedy. Examine primer of shell. If primer is deeply indented, faulty ammunition is indicated; if lightly indented, damaged firing mechanism is indicated and gun should be turned over to ordnance personnel for correction.

(3) Uncocked Hammer.

(a) Due to insufficient retraction of operating mechanism to cock hammer when loading shell into chamber, broken hammer spring, broken, bent, or incorrectly assembled hammer bar (Remington M31 and Ithaca M37 Guns), broken or burred sear or broken sear spring, foreign matter in sear notch of hammer or broken sear or trigger spring. In autoloading guns, incorrect assembly of friction ring and piece may cause trouble (pars. 58 and 67).

(b) To Remedy. Test functioning of operating mechanism, firing mechanism. Turn gun over to ordnance personnel for correction.

(4) Safety Engaged.

(a) Due to failure to disengage safety which will block trigger when pulled, thus preventing release of hammer, foreign matter in safety aperture or broken parts.

(b) To Remedy. Test positioning and retention of safety; turn gun over to ordnance personnel for correction.

(5) Unlocked Breech Bolt.

(a) Due to failure to move slide smartly and fully forward, obstruction in shell seat or extractor cuts in barrel, obstruction in top of receiver, swollen shell preventing full seating in chamber, or slide lock failing to hold slide in position.

(b) To Remedy. Push slide smartly and fully forward; reciprocate slide handle and relock bolt; examine above parts for obstruction or foreign matter; check slide lock for function. If breech bolt does not lock every time after removal of foreign matter, turn gun over to ordnance personnel for correction.

f. Failure to Load. Failure of the mechanism to load a shell into the chamber of the barrel may be due to a number of causes as follows:

(1) Blocked Chamber.

(a) Due to unextracted shell, or ruptured shell casing lodged in chamber.

(b) To Remedy. Place safety at safe position. Remove shells from

SHOTGUNS, ALL TYPES

magazine and receiver; close and lock action; then, retract operating handle to extract fired shell or remove (par. 73); examine chamber and bore, and remove broken casing or obstruction if present.

(2) FAULTY EXTRACTION AND EJECTION.

(a) Due to malfunction of or broken extractors or ejector, thus allowing unejected, fired shell to remain in chamber or receiver, thus blocking feeding of next live shell. In the Remington M11 and Sportsman, and the Savage M720 Guns, this fault may be due to improper assembly of friction ring and piece on magazine tube causing failure to eject (pars. 58 and 67).

(b) To Remedy. Remove fired shell and examine ejector and extractors for malfunction or fracture. Check position of friction ring and piece on magazine tube (autoloading guns) for load used. Examine ejector spring of Remington Gun M10 for tension (par. 42 *c* (9) and (12)).

(3) MALFUNCTIONS OF CARRIER.

(a) Due to damage, incorrect assembly, or presence of foreign matter, thereby preventing its proper functioning and alinement of shell being fed.

(b) To Remedy. Examine and test function of carrier and spring. Turn gun over to ordnance personnel for correction.

(4) DOUBLE FEEDING.

(a) Usually caused by malfunction of shell stop (or carrier latch, autoloading guns), thus allowing two shells to slip out of magazine into receiver during feeding operation. May be caused by foreign matter under shell stop or damaged stop.

(b) To Remedy. Examine and check function of stop (or latch); clean and adjust. If damaged, turn gun over to ordnance personnel for correction.

g. Failure to Feed. Failure to feed a shell into position to be loaded into the chamber may be due to blocking through failure of loading function or other causes as follows:

(1) JAMMED MAGAZINE.

(a) Due to broken or kinked magazine spring, dented or fouled tube, bent or fouled follower or swollen shell.

(b) To Remedy. Move magazine follower back and forth in magazine to loosen. If action of follower is not smooth and free, turn gun over to ordnance personnel for correction.

(2) MALFUNCTION OF CARRIER. Same as in f (3) above.

(3) MALFUNCTION OF SHELL STOPS.

(a) Due to foreign matter under stop, broken or deformed stop, or damaged or missing stop spring.

STOPPAGES AND IMMEDIATE ACTION

(b) To Remedy. Check function of stop; clean if fouled. If damaged, turn gun over to ordnance personnel for correction.

h. Failure of Action Slide Handle to Retract (Slide Action Guns). Failure of the action slide handle to retract may be due to several reasons:

(1) FAILURE OF ACTION SLIDE LOCK TO DISENGAGE WHEN MANUALLY OPERATED OR WHEN GUN IS FIRED.

(a) Due to improper operation, burs on slide lock, foreign matter in assembly, or improper assembly or broken parts.

(b) To Remedy. Check function of lock, clean mechanism. If still faulty, turn gun over to ordnance personnel for correction.

(2) FIRED SHELL STUCK IN CHAMBER.

(a) Due to swollen shell or fouling in chamber, preventing extraction by usual methods.

(b) To Remedy. Push safety to safe position and unload magazine. If extractors are visible, disengage from shell if possible; retract slide, unload magazine, and remove shell in chamber (par. 73).

(3) JAMMED HAMMER, SLIDE OR BOLT.

(a) Due to foreign matter in mechanism, broken hammer, or hammer spring, disengaged or broken hammer bar (Ithaca M37 or Remington M31 Guns), or other part of firing mechanism.

(b) To Remedy. Slide safety to safe position; fully unload gun; remove trigger plate group. Inspect and clean mechanism. If damaged, turn gun over to ordnance personnel for correction.

(4) JAMMED ACTION SLIDE BAR.

(a) Due to bent or twisted bar, causing excessive friction with receiver.

(b) To Remedy. Turn gun over to ordnance personnel for correction.

(5) JAMMED ACTION SLIDE LOCK.

(a) Due to deformed or burred lock mechanism or foreign matter in assembly, thus preventing disengagement of lock from slide or action slide bar, and preventing retraction of slide handle.

(b) To remedy. Slide safety to safe position. Fully unload gun. Check function of lock; remove trigger plate group (or operating mechanism); examine and clean lock mechanism. If still faulty, turn gun over to ordnance personnel for correction.

i. Safety Fails to Operate.

(1) May be due to burs on safety or housing, or foreign matter blocking safety, or missing or broken parts.

(2) TO REMEDY. Remove trigger plate group. Check operation of safety. Clean and lubricate as far as possible. If still inoperative, turn gun over to ordnance personnel for correction.

j. Trigger Fails to Pull or Release.

SHOTGUNS, ALL TYPES

(1) The failure of the trigger to pull may be due to incorrectly positioned safety, burs, damaged or missing parts, or foreign matter in mechanism. If trigger fails to release (move forward) before pulling, it may be due to above reasons or broken or improperly assembled trigger spring.

(2) To REMEDY. Remove trigger plate group. Check function of safety and trigger and clean assembly. If damaged, turn over to ordnance personnel for correction.

k. Breech Bolt Fails to Lock.

(1) This may be caused by foreign matter in the extractor cuts in receiver or barrel or in the locking aperture in top of the receiver (or barrel extension, autoloading guns), preventing proper seating of the breech bolt. Again the breech bolt may slip from the locked position due to failure of the slide lock to function to block the slide and, hence, the breech bolt in position. It also may be due to incompletely seated shell due to foreign matter in chamber seat or bore, or swollen shell, burs or broken firing pin (Stevens slide action, Remington or Savage autoloading Guns.)

(2) To REMEDY. Check function of parts; remove group from guns. Inspect, clean, and test. If shell causes trouble, place safety in safe position, extract shell and unload gun before inspection. Inspect bore and shell seat in barrel. If due to faulty parts, turn gun over to ordnance personnel for correction.

l. Breech Bolt Fails to Unlock.

(1) Failure to unlock may be due to jammed shell, jammed slide, or other mechanism, or to failure of the slide lock when disengaged to release the slide or action slide handle bar. May be due to broken firing pin in Stevens slide action guns, and in Remington and Savage autoloading guns.

(2) To REMEDY. Work mechanism gently to ascertain trouble. If live shell is in chamber, place safety at safe, unload magazine, remove trigger plate, and remove operating groups, if possible. If this is not easily accomplished, turn gun over to ordnance personnel for correction.

m. Shell Fails to Fire.

(1) May be due to faulty ammunition, broken or jammed firing pin, broken or weak mainspring (par. e (2) *(b)* above and sec. XIII).

(2) To REMEDY. Examine primer of shell; if lightly dented, broken or jammed parts are indicated as above; if normally dented, faulty ammunition is indicated. Remove and inspect breech bolt and trigger plate group. Turn gun over to ordnance personnel for correction.

n. Gun Fires as Breech Bolt Locks.

(1) May be due to pressure kept on trigger by operator as action slide handle is pushed forward, failure of sear to retain hammer due to

STOPPAGES AND IMMEDIATE ACTION

burs, foreign matter, or broken parts. Mechanism of Remington M31, M11, and Sportsman, and Savage M720 Guns is designed to make this fault impossible, unless parts are damaged or broken. Trigger must be released and again pulled before hammer is released.

(2) To REMEDY. Unload and test functioning of gun; remove trigger plate group; inspect parts and clean. If gun still fails to function properly, turn over to ordnance personnel for correction.

NOTE: Refer to paragraph 97 for causes of shell failure.

73. TO REMOVE A SHELL FROM THE CHAMBER.

a. If a fired shell becomes jammed in the chamber of the barrel so that it cannot be extracted in the usual manner and the extractors are engaged, unload the magazine and insert a cleaning rod in the muzzle end of the barrel. Then disengage the action slide lock and push shell out of chamber at the same time retracting the action slide handle to unlock the bolt.

CAUTION: Be sure shell has been fired before attempting removal.

b. If a live shell becomes jammed in the chamber and cannot be extracted in the usual manner, proceed as follows:

(1) Slide safety to the safe position.

(2) Unload magazine.

(3) Turn gun over to officer, qualified mechanic, or ordnance personnel for correction.

CAUTION: Do not attempt to ram a live shell from the chamber even if breechbolt is retracted, as primer may strike bolt and fire shell.

74. TO REMOVE A SHELL FROM THE MAGAZINE.

a. If a shell becomes jammed in the magazine so that the force of the magazine spring will not push it out into the receiver when the shell stop is disengaged from the base of the shell, proceed as follows:

(1) Slide safety to safe position and unload chamber.

(2) Using the finger or blunt stick, attempt to work shell gently back and forth in magazine to free it.

CAUTION: Do not use sharp instrument or strike base of shell, as detonation of the primer may result and the shell be fired.

(3) If shell cannot be freed in the manner prescribed in *(a)* above turn gun over to qualified ordnance personnel for correction.

(4) If shell can be removed, examine shell, magazine tube, and follower. If shell is swollen or deformed, fault is probably with the shell. If, after shell is removed, the follower still fails to move smoothly the full distance in the tube or tube follower appear dented or corroded, turn gun over to qualified ordnance personnel for correction.

SHOTGUNS, ALL TYPES

Section XI

SPECIAL MAINTENANCE

	Paragraph
Care and cleaning in cold climates	75
Care and cleaning in hot climates	76
Preparation of shotguns for storage	77
Cleaning of shotguns as received from storage	78

75. CARE AND CLEANING IN COLD CLIMATES.

a. In temperatures below freezing, it is necessary that the moving parts of the shotgun be kept absolutely free from moisture. It has also been found that excess oil on the working parts will solidify to such an extent as to cause sluggish operation or complete failure.

b. The shotgun should be disassembled under supervision of an officer or the chief mechanic and metal parts completely cleaned with SOLVENT, dry-cleaning, before use in temperatures below 0 F. The working surfaces of parts which show signs of wear may be lubricated by rubbing with a cloth which has been wet with OIL, lubricating, preservative, light, and wrung out. At temperatures above 0 F, the shotgun may be oiled lightly after cleaning by wiping with a slightly oiled cloth, using OIL, lubricating, preservative, light.

c. Upon bringing indoors, the shotgun should be allowed first to come to room temperature. It should then be disassembled, wiped completely dry of the moisture which will have condensed on the cold metal surfaces, and thoroughly oiled with OIL, lubricating, preservative, light. If possible, condensation should be avoided by providing a cold place in which to keep shotguns when not in use. For example, a separate cold room with appropriate racks may be used, or, when in the field, racks under proper cover may be set up outdoors.

d. If shotgun has been fired, it should be thoroughly cleaned and oiled. The bore may be swabbed out with an oily patch, and when the weapon reaches room temperature, thoroughly cleaned and oiled as prescribed in paragraph 11 and paragraphs pertaining to the cleaning of the gun in question.

e. Before firing, the shotgun should be cleaned and oil removed (par. b above). The bore and chamber should be entirely free of oil before firing.

76. CARE AND CLEANING IN HOT CLIMATES.

a. **Tropical Climates.**

(1) In tropical climates where temperature and humidity are high or where salt air is present, and during rainy seasons, the shotgun should

SPECIAL MAINTENANCE

be thoroughly inspected daily and kept lightly oiled when not in use. The groups should be dismounted at regular intervals and, if necessary, disassembled under the supervision of an officer or the chief mechanic, sufficiently to enable the drying and oiling of parts.

(2) Care should be exercised to see that unexposed parts and surfaces are kept clean and oiled, such as the under side of the barrel, magazine tube, inner surface of slide handle tube, interior of magazine tube, slide handle bar, interior of receiver, operating parts, trigger group, spring wells, and similar parts and surfaces.

(3) OIL, lubricating, preservative, light, should be used for lubrication.

(4) Wood parts should also be inspected to see that swelling due to moisture does not bind working parts. (If swelling has occurred, shave off only enough wood to relieve binding.) A light coat of OIL, linseed, raw, applied at intervals and well rubbed in with the heel of the hand will help to keep moisture out. Allow oil to soak in for a few hours and then wipe and polish with dry, clean rag.

NOTE: Care should be taken that linseed oil does not get into mechanism or on metal parts as it will become gummy when dry. Varnished surfaces will not absorb linseed oil.

b. **Hot, Dry Climates.**

(1) In hot, dry climates where sand and dust are apt to get into the mechanism and bore, the shotgun should be wiped clean daily or oftener, if necessary. Groups should be dismounted, and disassembled under the supervision of an officer or the chief mechanic, as far as necessary to facilitate thorough cleaning.

(2) When the shotgun is being used under sandy conditions, all lubricant should be wiped from the weapon. This will prevent sand carried by the wind from sticking to the lubricant and forming an abrasive compound which will ruin the mechanism. Immediately upon leaving sandy terrain, the weapon must be relubricated with OIL, lubricating, preservative, light.

(3) In such climates wood parts are apt to dry out and shrink. A light application of OIL, linseed, raw, applied as in a (4) above, will help to keep wood in condition.

(4) Perspiration from the hands is a contributing factor to rust because it contains acid. Therefore, metal parts should be wiped dry frequently.

(5) During sand or dust storms, breech and muzzle should be kept covered, if possible.

77. PREPARATION OF SHOTGUNS FOR STORAGE.

a. OIL, lubricating, preservative, light, is the most suitable oil for preserving the mechanism of shotguns. This oil is efficient for preserving

SHOTGUNS, ALL TYPES

the polished surfaces, the bore, and the chamber for a period of from two to six weeks, dependent on the climatic and storage conditions. Shotguns in short term storage should be inspected every five days and the preservative film renewed if necessary.

b. COMPOUND, rust-preventive, light, is a semisolid material. This compound is efficient for preserving the polished metal surfaces, the bore, and the chamber for a period of one year or less, dependent on the climatic and storage conditions.

c. Shotguns should be cleaned and prepared with particular care. The bore, all parts of the mechanism, and the exterior of the shotguns should be thoroughly cleaned and then dried completely with rags. In damp climates, particular care must be taken to see that the rags are dry. After drying a metal part, the bare hands should not touch that part. All metal parts should then be coated either with OIL, lubricating, preservative, light, or COMPOUND, rust-preventive, depending on the length of storage (a and b above). Application of the COMPOUND, rust-preventive, to the bore of the shotgun is best done by dipping the cleaning brush or patch assembled to cleaning rod in COMPOUND, rust-preventive, and running it through the bore two or three times. Cleaning brush must be clean. Before placing the shotgun in the packing chest see that the bolt is in its forward position and that the hammer is released to relieve tension of spring. Then, handling the shotgun by the wood parts only, it should be placed in the packing chest, the wooden supports at the butt and muzzle having previously been painted with COMPOUND, rust-preventive. Under no circumstances should a shotgun be placed in storage contained in a cloth or other cover or with a plug in the bore. Such articles collect moisture which causes the weapon to rust.

NOTE: If shotguns are packed in cardboard containers or original manufacturer's cartons, each gun or section of gun, if taken-down, should be wrapped in oilproof paper after preparing for storage.

78. CLEANING OF SHOTGUNS AS RECEIVED FROM STORAGE.

a. Shotguns which have been stored in accordance with paragraph 77 above will be coated with either OIL, lubricating, preservative, light, or COMPOUND, rust-preventive, light. Shotguns received from ordnance storage will, in general, be coated with COMPOUND, rust-preventive, heavy. Shotguns received from manufacturers will usually be coated with "Cosmoline," which is similar to COMPOUND, rust-preventive, light. Remove as much of compound and oil as possible with rags; then, use SOLVENT, dry-cleaning, to remove all traces of the compound, or oil. Particular care should be taken to see that all recesses in which springs or plungers operate, trigger mechanism and slide lock groups (or similar parts) are cleaned thoroughly (par. 11). After using the

SPECIAL MAINTENANCE

SOLVENT, dry-cleaning, make sure it is completely removed from all parts with clean, dry rags. Then oil and lubricate gun (par. 11). If guns are to be used immediately, bore and chamber should be wiped dry of oil and excess oil removed from exterior surfaces with a dry rag.

CAUTION: Failure to clean the firing pin and the recess in the bolt in which it operates, or the slide lock and trigger mechanism, may result in gun failure at normal temperatures, and will most certainly result in serious malfunctions if the shotguns are operated in low temperature areas, as COMPOUND, rust-preventive, and lubricating oil will congeal or frost on the mechanism.

b. SOLVENT, dry-cleaning, is an inflammable noncorrosive petroleum distillate, of low inflammability, used for removing grease. It should not be used near open flame, and smoking is prohibited where it is being used. It is generally applied with rag swabs to large parts and as a bath for small parts. The surfaces must be thoroughly dried immediately after removal of the solvent. To avoid leaving finger marks, which are ordinarily acid and induce corrosion, gloves should be worn by persons handling parts after such cleaning. SOLVENT, dry-cleaning, will attack and discolor rubber.

SHOTGUNS, ALL TYPES

Section XII

MATERIEL AFFECTED BY GAS

	Paragraph
Protective measures	79
Cleaning	80
Decontamination	81
Special points pertaining to shotguns	82

79. PROTECTIVE MEASURES.

a. When materiel is in constant danger of gas attack, unpainted metal parts will be lightly coated with engine oil. Instruments are included among the items to be protected by oil from chemical clouds or chemical shells, but ammunition is excluded. Care will be taken that the oil does not touch the optical parts of instruments or leather or canvas fittings. Materiel not in use will be protected with covers as far as possible. Ammunition will be kept in sealed containers.

b. Ordinary fabrics offer practically no protection against mustard gas or lewisite. Rubber and oilcloth, for example, will be penetrated within a short time. The longer the period during which they are exposed, the greater the danger of wearing these articles. Rubber boots worn in an area contaminated with mustard gas may offer a grave danger to men who wear them several days after the bombardment. Impermeable clothing will resist penetration more than an hour, but should not be worn longer than this.

80. CLEANING.

a. All unpainted metal parts of materiel that have been exposed to any gas except mustard and lewisite must be cleaned as soon as possible with **SOLVENT**, dry-cleaning, or **ALCOHOL**, denatured, and wiped dry. All parts should then be coated with engine oil.

b. Ammunition which has been exposed to gas must be thoroughly cleaned before it can be fired. To clean ammunition use **AGENT**, decontaminating, noncorrosive, or if this is not available, strong soap and cool water. After cleaning, wipe all ammunition dry with clean rags. *Do not use dry powdered AGENT, decontaminating (chloride of lime used for decontaminating certain types of materiel) on or near ammunition supplies,* as flaming occurs through the use of chloride of lime on liquid mustard.

81. DECONTAMINATION.

a. For the removal of liquid chemicals (mustard, lewisite, etc.) from materiel, the following steps should be taken:

MATERIEL AFFECTED BY GAS

(1) Protective Measures.

(a) For all of these operations a complete suit of impermeable clothing and a service gas mask will be worn. Immediately after removal of the suit, a thorough bath with soap and water (preferably hot) must be taken. If any skin areas have come in contact with mustard, if even a very small drop of mustard gets into the eye, or if the vapor of mustard has been inhaled, it is imperative that complete first-aid measures be given within 20 to 30 minutes after exposure. First-aid instructions are given in **TM 9-850** and **FM 21-40**.

(b) Garments exposed to mustard will be decontaminated. If the impermeable clothing has been exposed to vapor only, it may be decontaminated by hanging in the open air, preferably in sunlight for several days. It may also be cleaned by steaming for two hours. If the impermeable clothing has been contaminated with liquid mustard, steaming for six to eight hours will be required. Various kinds of steaming devices can be improvised from materials available in the field.

(2) Procedure.

(a) Commence by freeing materiel of dirt through the use of sticks, rags, etc., which must be burned or buried immediately after this operation.

(b) If the surface of the materiel is coated with grease or heavy oil, this grease or oil should be removed before decontamination is begun. **SOLVENT**, dry-cleaning, or other available solvents for oil should be used with rags attached to ends of sticks.

(c) Decontaminate the painted surfaces of the materiel with bleaching solution made by mixing one part **AGENT**, decontaminating (chloride of lime), with one part water. This solution should be swabbed over all surfaces. Wash off thoroughly with water, then dry and oil all surfaces.

(d) All unpainted metal parts and instruments exposed to mustard or lewisite must be decontaminated with **AGENT**, decontaminating, noncorrosive, mixed one part solid to fifteen parts solvent (**ACETYLENE TETRACHLORIDE**). If this is not available, use warm water and soap. Bleaching solution must not be used, because of its corrosive action. Instrument lenses may be cleaned only with **PAPER**, lens, tissue, using a small amount of **ALCOHOL**, ethyl. Coat all metal surfaces lightly with engine oil.

(e) In the event **AGENT**, decontaminating (chloride of lime), is not available, materiel may be temporarily cleaned with large volumes of hot water. However, mustard lying in joints or in leather or canvas webbing is not removed by this procedure and will remain a constant source of danger until the materiel can be properly decontaminated. All mustard washed from materiel in this manner lies unchanged on the

SHOTGUNS, ALL TYPES

ground, necessitating that the contaminated area be plainly marked with warning signs before abandonment.

(f) The cleaning or decontaminating of materiel contaminated with lewisite will wash arsenic compounds into the soil, poisoning any water supplies in the locality for either men or animals.

(g) Leather or canvas webbing that has been contaminated should be scrubbed thoroughly with bleaching solution. In the event this treatment is insufficient, it may be necessary to burn or bury such materiel.

(h) Detailed information on decontamination is contained in FM 21-40, TM 9-850, and TC 38, 1941, Decontamination.

82. SPECIAL POINTS PERTAINING TO SHOTGUNS.

a. The shotguns should be completely disassembled under the supervision of an officer or the chief mechanic, for cleaning and decontamination, and special attention given to the points enumerated below:

(1) Bore and chamber.

(2) Interior of magazine tube and parts.

(3) Guideways and apertures in receiver.

(4) Trigger mechanism.

(5) Inner surface of action slide handle tube or like part.

(6) Apertures in breech bolt or like parts.

(7) Spring wells.

(8) Friction parts and inside of fore end (autoloading guns).

(9) Apertures in carrier of Winchester Gun M97.

(10) Action spring tube and spring (autoloading guns).

(11) Bayonet attachment, hand guard, and sling.

(12) Safety apertures.

(13) Bayonet and bayonet scabbard, with special attention to catch, and interior of scabbard.

(14) Gun sling. (Disassemble when cleaning and decontaminating).

Section XIII

AMMUNITION

	Paragraph
General	83
Nomenclature	84
Classification	85
Firing tables	86
Identification	87
Lot number	88
Grade	89
Marking	90
Packing and marking	91
Care, handling, and preservation	92
Precautions in firing	93
Storage	94
Authorized rounds	95
Data	96
Defects found after firing	97
Field report of accidents	98

83. GENERAL.

a. The information in this section pertaining to the several types of shotgun shells (shot shells) authorized for use in authorized commercial standard shotguns of pump action and autoloading types, includes description, means of identification, care, use, and ballistic data. Shotgun shells are procured by the Ordnance Department from several commercial manufacturers.

84. NOMENCLATURE.

a. Standard nomenclature is used herein in all references to specific items of issue.

85. CLASSIFICATION.

a. Based upon use, the shotgun shells used in these shotguns may be classified as:

(1) Guard or combat load—contains $1\frac{1}{8}$ ounces of No. 00 buckshot.

(2) Trap load—contains $1\frac{1}{8}$ or $1\frac{1}{4}$ ounces of No. $7\frac{1}{2}$ chilled shot.

(3) Skeet load—contains $1\frac{1}{8}$ ounces of either No. $7\frac{1}{2}$ chilled shot or No. 9 chilled shot.

(4) Hunting load—contains any of the above mentioned loads.

SHOTGUNS, ALL TYPES

86. FIRING TABLES.

a. Ballistic data are published herein in paragraph 96. They will not be published separately.

87. IDENTIFICATION.

a. General. Since shotgun shells are commercial items, they do not have any model designation. However, although they all have the same general appearance, each type of shotgun shell may be identified by the stamping on the metal head, shell body, and closing wad (par. 90) and by the markings on packing containers and boxes (par. 91).

b. Description. Shotgun shells consist of a brass or steel head, a primer, a paper case or shell body, a smokeless powder propelling charge, cardboard and felt wads, a load of lead shot, and a closing wad. In addition to the markings on the ammunition and packing containers, the shotgun shells may be further identified by the physical characteristics described below:

(1) SHOTGUN SHELL FOR GUARD OR COMBAT USE. These shells have waterproofed paper cases, and metal heads, ranging from approximately 1 inch down to .35 inch in length (fig. 127). Metal heads approximately .80 inch in length or above will only appear on shells having a guard or combat load, although many of these are now equipped with the low (.35) metal heads. In the past some shells have had the entire case made of brass and these are authorized for guard or combat use only.

(2) SHOTGUN SHELL FOR TRAP OR SKEET SHOOTING. These consist of a paper case or shell body set in a metal head which is under ½ inch in length.

(3) SHOTGUN SHELLS FOR HUNTING. Any of the above shells, with the exception of those having the all-brass case for guard or combat use only, may be used for the appropriate type of hunting.

88. LOT NUMBER.

a. When ammunition is manufactured, an ammunition lot number which becomes an essential part of the marking is assigned in accordance with pertinent specifications. This lot number is marked on all packing containers. It is required for all purposes of record, including grading, use, and reports on condition, functioning, and accidents in which the ammunition might be involved. Since it is impracticable to mark the ammunition lot number on each individual shotgun shell, every effort should be made to maintain the ammunition lot number of shells that have been removed from their original packings. Shotgun shells for which the ammunition lot number has been lost are placed in grade 3 (unserviceable ammunition which will not be fired). Therefore, when shells are removed from their original packings, they should be marked or tagged so that the ammunition lot number may be preserved.

AMMUNITION

89. GRADE.

a. No grades are assigned to serviceable shotgun shells. Grade 3 indicates unserviceable ammunition, which will not be fired but will be destroyed locally. Instructions for the destruction of ammunition are contained in TM 9-1900.

90. MARKING.

a. **General.** Stampings on the closing wad, paper case or shell body, and metal head of the shotgun shells (fig. 127) are used as a means of identification of the ammunition. Further identification is obtained from the markings on packing containers (par. 91) which includes that of the lot number (pars. 87 and 88).

b. **Marking on Closing Wad.**

(1) Markings stamped on closing wads are given in Table I. The numerals indicate the quantities of propellent and weight and size of the shot load, for example, "3—1⅛—7½C" indicates 3 drams equivalent of smokeless powder and 1⅛ ounces of No. 7½ chilled shot.

Table I.

Stamping on closing wad	Load
3—00—BUCK	Guard or combat
3—1⅛—7½ C	Trap or skeet
3—1¼—7½ C	Trap
3—1⅛—9 C	Skeet

(2) In addition, the name or symbol of the manufacturer and the name of the powder manufacturer may also be stamped on the closing wad.

c. **Marking on Case.** Some shotgun shells have the information stamped on the closing wad repeated on the paper case or shell body (b above). In addition, the trade name and type of load of the shotgun shell may also be stamped on the case.

d. **Marking on Metal Head.** The stampings on the metal head are indicated in Table II. This stamping generally consists of initials or symbol of manufacturer, gage size of shell, and trade name for the particular type of shotgun shell.

SHOTGUNS, ALL TYPES

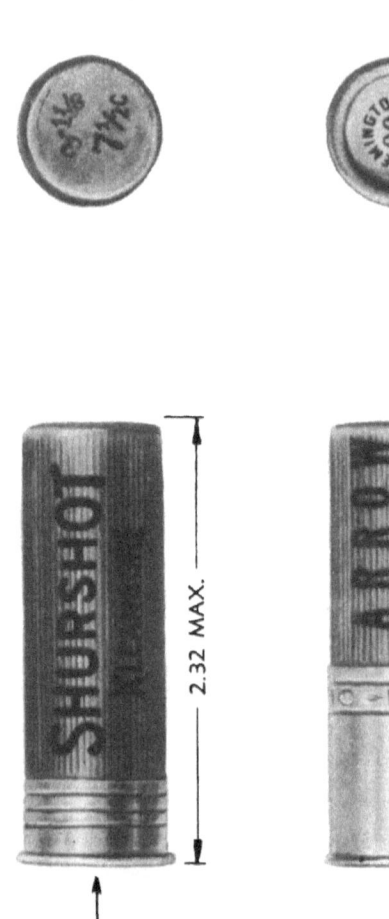

A — SHELL, SHOTGUN, 12-GAGE, PAPER, LOADED WITH SMOKELESS POWDER AND NO. 7-1/2 CHILLED SHOT
B — SHELL, SHOTGUN, 12-GAGE, PAPER, LOADED WITH SMOKELESS POWDER AND NO. 00 BUCKSHOT

Figure 127 — Shotgun Shells

AMMUNITION

Table II.

Manufacturer	Marking on metal head for:		
	Guard or combat load		Sporting load
	Former marking*	New marking**	
Federal Cartridge Co.	Federal No. 12 Hi Power	Federal No. 12 Hi Power	Federal No. 12 Monark
Peters Cartridge Co.	P-Peters No. 12 DeLuxe Target	P-Peters No. 12 High Velocity	P-Peters No. 12 Victor Trap Load
Remington Arms Co.	Rem UMC No. 12 Arrow	Rem UMC No. 12 Nitro Express	Rem UMC No. 12 Shurshot
Western Cartridge Co.	Western No. 12 Record	Western No. 12 Super X	Western No. 12 Xpert
Winchester Repeating Arms Co.	W.R.A. 12 Leader	W.R.A. 12 Superspeed	W.R.A. 12 Ranger

* Shotgun shell procured in accordance with U. S. specification 50-5-1A, February 17, 1939.
** Shotgun shell procured in accordance with U. S. specification 50-5-1B, April 28, 1942.

91. PACKING AND MARKING.

a. Packing. Shotgun shells are packed 25 shells per carton, 20 cartons (500 shells) per case. The case for domestic shipment may be wood or fiber; for oversea shipment, it will be metal-lined wood. Representative dimensions and weight are:

Size in inches .... 15 x 10⅜ x 9¾
Area .... 1.08 sq ft
Volume .... 0.88 cu ft
Weight .... 65 lb

b. Marking. Cases and cartons of this ammunition bear the commercial markings of the manufacturer and in addition the lot number, type of load, and the phrase "U. S. property." These markings generally include manufacturer's name and address, quantity, gage size, gun chamber length, type of ammunition, type of propellent powder, and such trade names as indicated in Table II. The phrases "Loaded shot shells" or "Small-arms ammunition" also appear on the packing cases.

92. CARE, HANDLING, AND PRESERVATION.

a. Small-arms ammunition, as compared with other types of ammunition, is not dangerous to handle. Care, however, must be observed to keep the packing cases from becoming broken or damaged. All broken cases must be immediately repaired and careful attention given to the transfer of all markings to the new parts of the box. In case the packing box contains a metal liner, the liner should be air tested and sealed provided that equipment for this work is available.

SHOTGUNS, ALL TYPES

b. Ammunition boxes should not be opened until the ammunition is required for use. Ammunition removed from its container, particularly in damp climates, is apt to corrode, thereby causing the ammunition to become unserviceable.

c. The ammunition should be protected from mud, sand, dirt, and water. If it gets wet or dirty, it should be wiped off at once. Verdigris or light corrosion, if it forms on the metal part of the shells, should be wiped off. However, shells should not be polished to make them look better or brighter. Ammunition which is seriously corroded should not be fired.

d. The use of oil or grease on cartridges is prohibited.

e. Do not fire shotgun shells with defects.

f. Ammunition should not be exposed to the direct rays of the sun for any length of time. Such exposure may affect seriously its firing qualities.

g. Whenever shotgun shells are taken from original packing containers, they will be so tagged or otherwise marked so that the ammunition may be identified as to lot number. Such identification is necessary to prevent otherwise serviceable ammunition from being placed in grade 3, because of loss of lot number.

93. PRECAUTIONS IN FIRING.

a. Because a misfire cannot immediately be distinguished from a hangfire, when a misfire occurs, the ejection opening of the gun should be turned toward the ground and the shotgun shell ejected. Should the malfunction be a hangfire, this procedure will prevent the resulting flame from causing injury to the firer. Although some models of shotguns can be recocked and refired, for example the Winchester M97, the above procedure is strongly recommended should a misfire occur.

b. Before firing, the firer should be sure that the bore of the weapon is free of any foreign matter such as cleaning patches, mud, sand, snow, and the like, and especially a shell of smaller gage which has been placed or lodged in the barrel through error. To fire a weapon with any obstruction in the bore may cause the gun to burst and result in injury to the firer.

c. Shotgun shells should not be fired unless they are identified by lot number. Such identification prevents the ammunition from becoming grade 3 ammunition.

d. It should be kept in mind that the shotgun shells issued by the Ordnance Department are designed for use in authorized commercial standard shotguns having a chamber $2\frac{3}{4}$ inches in length. Although

AMMUNITION

the occasion may arise whereby this ammunition is used in shotguns having longer chambers, the shotgun shells should not be used in shotguns having chambers shorter than 2¾ inches. To do so may result in dangerously high chamber pressures.

94. STORAGE.

a. Whenever practicable, small-arms ammunition should be stored under cover. When necessary to have small-arms ammunition in the open, it should be raised at least six inches from the ground and covered with a double thickness of tarpaulin. Suitable trenches should be dug to prevent water from flowing under the pile.

b. If placed in a fire, small-arms ammunition does not explode violently. There are small individual explosions of each shell, the case flying in one direction and the shot in another. In case of fire, keep personnel not engaged in fighting the fire at least 200 yards from the fire and have them lie on the ground. It is unlikely that the shot and cases will fly over 200 yards.

c. Small-arms ammunition in storage should be protected from extreme heat in order to avoid decomposition of the propellent powder. The combination of high temperature and damp atmosphere is particularly detrimental to the stability of the powder.

d. When only a part of a box is used, the remaining ammunition in the box should be protected against unauthorized handling and use by firmly fastening the cover in place.

95. AUTHORIZED ROUNDS.

a. The following ammunition is authorized for use in authorized commercial standard 12-gage shotguns. It should be noted that the nomenclature completely describes the shell as to type and gage size. Its use for all purposes of record is mandatory.

SHELL, shotgun, 12-gage, brass, loaded with smokeless powder and ♯00 buckshot[1]

SHELL, shotgun, 12-gage, paper, loaded with smokeless powder and ♯00 buckshot[2]

SHELL, shotgun, 12-gage, paper, loaded with smokeless powder and ♯7½ chilled shot

SHELL, shotgun, 12-gage, paper, loaded with smokeless powder and ♯9 chilled shot[1]

96. DATA.

a. **General.** Although there are slight differences in shotgun shells of

SHOTGUNS, ALL TYPES

different manufacture (powder charge, etc.), the following data in Table III and Table IV are considered substantially correct.

Table III.

	Type of Load		
	Guard or combat	Trap	Skeet
Weight of bulk powder charge (equiv.), drams	3	3	3
Size of shot	No. 00 buck	No. 7½	No. 9
Weight of shot load, ounces	1⅛	1¼	1⅛
Number of shot in load	9	445±15	660±40
Length, inches	2.33	2.33	2.33
Weight of shotgun shell, grains	800	794	741
Maximum chamber pressure, pounds per square inch	11,000[6]	11,000[6]	11,000[7]
Average velocity over a range of 40 yards, feet per second	1,070±30[6]	850±30[6]	940±40[8]
Pattern in percentage of shot in a 30-inch circle	70% at 40 yards[6]	60% at 40 yards[6]	50% at 25 yards[7]
Maximum effective range, yards	60-75[6,7]	45-50[6] 30-35[7]	25-30[7]

[1] Guard or combat load. This item is issued only for this use.
[2] Guard or combat load.
[3] Trap or skeet load.
[4] Skeet load.
[5] For shotgun shell procured in accordance with specification 50-5-1A, February 17, 1939.
[6] Fired in 30-inch full choke barrel.
[7] Fired in 26-inch cylinder barrel.
[8] Fired in 26-inch cylinder or skeet bored barrel over a range of 25 yards. When fired in 30-inch full choke barrel over a range of 40 yards, the average velocity is 1,050±40 feet per second.

Table IV.

	Type of Load	
	Guard or combat	Trap or skeet
Weight of bulk powder charge (equiv.), drams	3	3
Size of shot	No. 00 buck	No. 7½
Weight of shot load, ounces	1⅛	1⅛
Number of shot in load	9	435±15
Length, inches	2.33	2.33
Weight of shotgun shell, grains	800	750
Maximum chamber pressure, pounds per square inch	11,000	11,000
Average velocity over range of 40 yards fired in 30-inch full choke barrel, feet per second	1,070±30	870±30
Pattern in percentage of shot in a 30-inch circle at 40 yards.	33⅓%[2]	65%[3] 40%[4]
Maximum effective range, yards	60-75[3,4]	45-50[3] 30-35[4]

[1] For shotgun shell procured in accordance with specification 50-5-1B, April 28, 1942.
[2] Fired in 20-inch cylinder bored, riot type barrel.
[3] Fired in 30-inch full choke barrel.
[4] Fired in 26-inch cylinder bored barrel.

AMMUNITION

b. **Table of Fire.** The following table indicates the pattern or dispersion as a percentage of the total number of shot falling within a circle of 30-inch diameter at the range indicated. The approximate pattern spread is also indicated. These values are only approximate since there is considerable variation in shotgun ballistics. This variation may be due not only to a particular loading but also to atmospheric conditions.

Table V.

Range, in yards	Guard or combat load[1]		Sporting trap,[2]	Sporting trap or sporting skeet load[3]		Sporting skeet load,[4] fired in 26-inch cylinder bore barrel
	Fired in 30-inch full choke barrel	Fired in 26-inch cylinder bore barrel	Fired in 30-inch full choke barrel	Fired in 26-inch cylinder or skeet bore barrel		
5	100%	100%	—	—		—
10	100%	100%	—	—		—
15	100%	100%	—	—		—
20	100%	100%	100%	75%		87%
25	100%	100%	100%	57%		66%
30	100%	100%	90%	45%		59%
35	90%	70%	80%	40%		43%
40	75%	60%	70%	33%		33%
45	—	—	—	—		—
50	50%	38%	—	—		—
55	—	—	—	—		—
60	35%	25%	—	—		—
Maximum effective range	60-75 yd	60-75 yd	45-50 yd	30-35 yd		25-30 yd
Pattern spread per yd	¾ in.	1 in.	9/10 in.	1⅛ in.		1⅛ in.

[1] Contains 9 (1⅛ oz) No. 00 buckshot.
[2] Contains 445±15 (1¼ oz) No. 7½ chilled shot.
[3] Contains 435±15 (1⅛ oz) No. 7½ chilled shot.
[4] Contains 660±40 (1⅛ oz) No. 9 chilled shot.

97. DEFECTS FOUND AFTER FIRING.

Name of defect	How to recognize	Common causes—precautions
Misfire	No action on firing. Primer shows normal impression of firing pin.	Primer is defective.
	No action on firing. Primer shows light impression of firing pin.	Indicates mechanical defect in weapon as short or broken firing pin, weak firing pin spring, bolt of weapon not being completely locked, or grease in firing pin hole which cushions blow of firing pin, or caused by defective shell or primer.
	No action on firing. Primer shows normal impression of firing pin, but off center.	Defect in weapon.

SHOTGUNS, ALL TYPES

Name of defect	How to recognize	Common causes—Precautions
Hangfire	Delayed ignition of powder in the shell.	Small or decomposed primer pellet, damp powder or light blow of firing pin caused by dirt or defect in weapon. This is a serious defect if delay is long enough to permit the bolt to be opened before the powder burns completely, in which case, injury to firer or damage to weapon, or both, may result. A hangfire in some shotguns may even force the breech open. For precautions, see paragraph 93.
Pierced primer	Perforation of primer cup by the firing pin. Discoloration around indent of very small perforation. Disk from large perforation blown into action of gun, with such an escape of gas as to lower velocity of the shot.	Imperfect firing pin or very thin metal in base of primer cup.
Primer leak	Discoloration around the primer and the metal head. Slight discoloration when primer leak is small, or heavy for a large primer leak.	Too small a primer, too large a primer hole, or excessive pressure generated by propelling charge.
Blown primer	Primer is blown completely from pocket of metal head or case.	Seldom encountered.
Primer setback	Primer protrudes above the metal head.	Defective bolt or shell or excessive pressure.
Leak back of case	Discoloration along the shell body or case.	Escape of gas into the action of weapon.
Failure of case to extract	Failure of case to extract.	Defective extractor or shell, a swollen case or dirty chamber.
Blowback	Escape of gas to the rear.	Pierced primer, primer leak, blown primer, and ruptured shell case.
Split body	Longitudinal split in shell body or case, thereby reducing velocity of the shot.	Case of body made of defective materiel.
Complete rupture	Circumferential separation completely around shell body or case, causing it to separate into two parts.	Bad bolt locking, excessive headspace, or defective shell case. This is a serious defect, because if the forward portion of the case remains in the chamber, it will cause the next round to jam.
Partial rupture	Partial circumferential separation around shell body or case.	See "Complete rupture."

98. FIELD REPORT OF ACCIDENTS.

a. Any serious malfunctions of ammunition must be reported promptly to the ordnance officer under whose supervision the materiel is maintained or issued (par. 7, AR 45-30).

Section XIV

REFERENCES

	Paragraph
Standard nomenclature lists	99
Explanatory publications	100

99. STANDARD NOMENCLATURE LISTS.

a. Cleaning, preserving and lubricating materials, recoil fluids, special oils, and similar items of issue. SNL K-1

b. Shotguns—parts, equipment and appendages..... SNL B-9

c. Shells, shotgun .... SNL T-3

d. Soldering, brazing and welding material, gases and related items .... SNL K-2

e. Tools, maintenance, for repair of small and hand arms and pyrotechnic projectors................ SNL B-20

f. Truck, small arms repair, M1—parts, equipment and load .... SNL G-72

Current Standard Nomenclature lists are as tabulated here. An up-to-date list of SNL's is maintained as the "Ordnance Publications for Supply Index".... OPSI

100. EXPLANATORY PUBLICATIONS.

a. **Ammunition.**

(1) Ammunition, general .... TM 9-1900
(2) Small-arms, ammunition .... TM 9-1990
(3) Qualifications in arms and ammunition training allowances .... AR 775-10

b. **Gas Attack.**

Defense against chemical attack................ FM 21-40
Decontamination, 1941 .... TC No. 38
Military chemistry and chemical agents......... TM 3-215

c. **Maintenance.**

Cleaning, preserving, lubricating and welding materials and similar items of issue by the Ordance Department .... TM 9-850
Ordnance Maintenance: shotguns, all types...... TM 9-1285
Ordnance maintenance procedure: materiel inspection and repair................................ TM 9-1100

TM 9-285
100

SHOTGUNS, ALL TYPES

d. **Ordnance Storage and Shipment.**
Marking shipments of ordnance supplies .... IOSSC-(b)
Ordnance storage and shipment chart—Group B—
 Major items .... OSSC-B
e. Ordnance field service in time of peace .... AR 45-30

f. **Shooting and Targets.**
Shotgun and skeet shooting .... TM 1-1100
Targets, target materials, and rifle range construction .... TM 9-855
g. Instruction guide: small arms data .... TM 9-2200

TM 9-285

INDEX

A

	Page No.
Accidents, ammunition, field report of	248
Action bar lock	
Remington, M10	117–119
Remington, M31	142
Action slide group (take-down gun), functioning, Winchester, M97	31–32
Action slide lock	
Winchester, M12	42
Winchester, M97	16
Ammunition	
authorized rounds	245
care, handling, and preservation	243–244
classification	239
cleaning gas exposed	236
data	245–247
field report of accidents	248
firing	
defects found after	247–248
precautions in	244–245
table	247
general information on	239
grade	241
identification	240
lot number	240
marking	241–243
nomenclature, standard	239
packing and marking	243
Assembled gun (*See* Testing assembled gun)	
Assembled parts, faulty	
Ithaca, M37	109
Remington, M10	134–135
Remington, M11 and Sportsman	189–191
Remington, M31	158
Savage, M720	220–221
Stevens, M620A, etc.	85–86
Winchester, M12	56–57
Winchester, M97	33–34
Authorized rounds	245
Autoloading gun	3

B

Barrel	
boring	4
inspection	7
Remington, M11 and Sportsman	163
Savage	195
Winchester	
solid frame gun	31
take-down gun	31–32

	Page No.
Bayonet attachment, Winchester, M97	
description	16
functioning	32
Bolt, malfunction, effect of	9
Bore, cleaning, Winchester, M97	34–35
Boring of the barrel	4
Breech bolt	
Remington, M11 and Sportsman	167
Savage	200
Winchester, M12	39
Winchester, M97	13
"Breech" defined	9
Breech mechanism, care of	9
Breechblock	
Ithaca, M37	90
Remington, M10	117
Remington, M31	140–142

C

Care, handling, and preservation of ammunition	243–244
Carrier	
Remington, M10	114–117
Remington, M11 and Sportsman	166
Savage	198
Winchester, M97	13
Carrier latch	
Remington, M11 and Sportsman	166
Savage	198
"Cartridge," word "shell" substituted for	3
Chamber, removal of shell from	231
Chemicals, liquid, removal of	236–238
"Choke" defined	4
Choking, how accomplished	4
Classification	
general	5–7
Ithaca, M37	88–90
Remington, M10	112–114
Remington, M11 and Sportsman	163
Remington, M31	137–140
Savage, M720	195
Stevens, M620A, etc.	62
Winchester, M12	38–39
Winchester, M97	13
Cleaning	
bore of solid frame guns	5
gas exposed materiel	236
Cleaning and lubrication	
Ithaca, M37	110–111
Remington, M10	136

251

TM 9-285

SHOTGUNS, ALL TYPES

C—Cont'd

	Page No.
Cleaning and lubrication—(Cont'd)	
Remington, M11 and Sportsman	191–192
Remington, M31	160
Savage, M720	222–223
Stevens, M620A, etc.	86–87
Winchester, M12	58
Winchester, M97	34–37
Climates, hot and cold, care and cleaning in	232–233
"Clockwise" defined	4
Cold climates, care and cleaning in	232
(See also Cleaning and lubrication)	
Construction (See Description)	
"Counterclockwise" defined	4

D

Data	
ammunition	245–247
Ithaca, M37	94
Remington, M10	119–120
Remington, M11 and Sportsman	170
Remington, M31	143
Savage, M720	202
Stevens, M620A, etc.	69
Winchester, M12	41
Winchester, M97	16–17
Decontamination of materiel exposed to gas	236–238
Description	
Ithaca, M37	88–93
Remington, M10	112–119
Remington, M11 and Sportsman	161–167
Remington, M31	137–140
Savage, M720	193–200
Stevens, M620A, etc.	59–68
Winchester, M12	38–41
Winchester, M97	10–16
Design, variations in	3
Disassembly and assembly, general explanation of	5
Diameters of bore of barrel	4

E

Ejector, Stevens, M620A, etc.	65
Extractor, Remington, M10	119

F

Failures (See Malfunctions, possible cause and correction)	
"Featherlight" applied to Ithaca shotgun	88

	Page No.
Field inspection	
Ithaca, M37	108–110
Remington, M10	133–136
Remington, M11 and Sportsman	188–191
Remington, M31	157–159
Savage, M720	219–222
Stevens, M620A, etc.	84–86
Winchester, M12	55–58
Winchester, M97	32–34
Field report of ammunition accidents	248
Firing ammunition	
defects found after	247–248
precautions in	244–245
table	247
Firing mechanism	
"hammerless" defined	4
Remington, M10	117
Firing table	247
Friction ring, use of	
Remington, M11 and Sportsman	186–188
Savage, M720	217–219
Functioning	
bolt and/or slide locking mechanisms, care of	9
Ithaca, M37	96–98
Remington, M10	122–126
Remington, M11 and Sportsman	175–178
Remington, M31	146–149
Savage, M720	207–210
Stevens, M620A, etc.	73–75
Winchester, M12	46–47
Winchester, M97	20–23

G

Gas, materiel affected by	236–238
cleaning	236
decontamination	236–238
protective measures	236
special points	238
Grade	
ammunition	241
definition of	6
Group(s)	
definition of	5
removal and replacement (See under names of gun)	
Gun sling, Winchester, M97	16

TM 9-285

INDEX

H

	Page No.
Hammer, Stevens, M620A, etc.	66
"Hammerless" defined	4
Hot climates, care and cleaning in	232–233

I

	Page No.
Identification marks	
Ithaca, M37	88
Remington, M10	112
Remington, M11 and Sportsman	161
Remington, M31	137
Savage, M720	193
Stevens, M620A, etc.	59
Winchester, M12	38
Winchester, M97	10
Identification of ammunition	240
Immediate action (*See* Stoppages and immediate action)	
Inspection of barrel	7
(*See also* Field inspection)	
Ithaca shotgun, M37	
cleaning and lubrication	110–111
data	94
description	88–90
general	90–93
field inspection	108–110
assembled parts, faulty	109
general	109–110
operation of gun	108–109
functioning	96–98
groups	
barrel and bayonet attachment	
removal	102
replacement	107
breechlock slide and carrier	
removal	102–106
replacement	106–107
trigger plate	
removal	102
replacement	107
operation (*See* Operation)	
standard nomenclature, use of	3

L

	Page No.
Liquid chemicals, removal of	236–238
Loading (*See* Operation)	
Lot number of ammunition	240
Lubrication (*See* Cleaning and lubrication)	

M

	Page No.
Magazine	
Ithaca, M37	92
Remington, M10	114
Remington, M31	142
removal of shell from	231
Winchester, M12	39
Winchester, M97	
description	16
functioning	31-32
replacement	31
Magazine tube	
Remington, M11 and Sportsman	163–166
Savage, M720	198
Stevens, M620A, etc.	62
Maintenance	232–235
care and cleaning in:	
cold climates	232
hot climates	232–233
storage	
cleaning guns received from	234–235
preparing guns for	233–234
Malfunction of bolt and/or slide lock, effect of	9
Malfunctions, possible cause and correction	226–231
failure:	
action slide handle	229
breech bolt	230–231
safety	229
shell	230, 247–248
to feed	228–229
to fire	226–227
trigger	229–230
general discussion of	226
Marking ammunition	241–243
Materiel affected by gas (*See* Gas, materiel affected by)	
Measuring choke in the barrel	4
Migratory Game Act	7
Model, definition of	6
"Muzzle" defined	8

N

	Page No.
Nomenclature for slide handle	4
Nomenclature, standard, use of	3
ammunition	239

253

TM 9-285

SHOTGUNS, ALL TYPES

O

	Page No.
Operation	
Ithaca	94–96
during field inspection	108–109
loading and unloading:	
magazine and gun	96
Remington, M10	120–122
during field inspection	133–134
loading and unloading:	
loading the chamber only	122
magazine and gun	122
Remington, M11 and Sportsman	170–174
during field inspection	188–189
explanation	170–174
loading:	
barrel only	174
from magazine	171–174
magazine	171
unloading:	
barrel without unloading magazine	174
gun	174
magazine	171
Remington, M31	143–146
during field inspection	157
loading and unloading:	
gun	145–146
loading the chamber only	146
magazine	145
Savage, M720	202–205
during field inspection	219–220
explanation of	202–203
loading:	
barrel	205
from the magazine	203–205
magazine	203
unloading	
barrel without unloading magazine	205
gun	205
magazine	203
Stevens, M620A, etc.	69–73
during field inspection	84–85
loading and unloading:	
gun and magazine	71
loading the chamber only	73
Winchester, M12	41–46
during field inspection	56
loading and unloading:	
gun	44–46
loading chamber only	46
magazine	44

	Page No.
Winchester, M97	17–20
during field inspection	32–34
loading and unloading:	
gun	20
magazine	17–20

P

Packing and marking ammunition	243
Precautions, safety	8–9
Primer, description and action of	7

R

Receiver	
Ithaca, M37	90
Remington, M10	114
Remington, M11 and Sportsman	166
Remington, M31	140
Savage, M720	198
Stevens, M620A, etc.	62–65
Winchester, M12	39
Winchester, M97	13
Recoil, Remington M11 and Sportsman	161
Remington shotgun, M10	
cleaning and lubrication	136
data	119–120
description	112–114
general	114–119
field inspection	133–136
assembled parts, faulty	134–135
general	135–136
operation of gun	133–134
functioning	122–126
groups	
barrel, magazine and action bar group (take-down gun)	
removal	126
replacement	133
breech bolt and link	
removal	183–184
replacement	184
butt stock	
removal	126
replacement	130
trigger guard	
removal	126–129
replacement	130
operation	120–122
Remington shotgun, M11 and Sportsman	
cleaning and lubrication	192
data	170

INDEX

R—Cont'd

	Page No.
Remington shotgun, M11 and Sportsman—(Cont'd)	
description	161–163
general	163–167
field inspection	188–191
assembled parts, faulty	189–191
general	191
operation of gun	188–189
friction ring, use of	186–188
functioning	175–178
groups	
action spring	
removal	183
replacement	185
barrel and fore end	
removal	178–183
replacement	185–186
breech bolt and link	
removal	183–184
replacement	184
butt stock	
removal	178–183
replacement	185–186
carrier	
removal	183
replacement	184–185
trigger plate	
removal	183
replacement	185
operation (See Operation)	
Remington shotgun, M31	
cleaning and lubrication	160
data	143
description	137–140
general	140–143
field inspection	
assembled parts, faulty	158
general	158–159
operation of gun	157
functioning	146–149
groups	
barrel	
removal	149
replacement	156
slide, breechblock, and carrier	
removal	149–152
replacement	152–156
trigger plate	
removal	149
replacement	156
operation (See Operation)	
Report of ammunition accidents	248

	Page No.
Riot gun, classification of	6–7
Rust and corrosion, removal, Winchester, M97	36

S

Safety precautions	8–9
Safety slide, Remington shotgun, M10	119
Safety, Stevens shotgun, M620A, etc.	68
Savage shotgun, M720	
cleaning and lubrication	222–223
data	202
description	193–195
field inspection	219–222
assembled parts, faulty	220–221
general	221–222
operation of gun	219–220
friction ring, use of	217–219
functioning	207–210
groups	
action spring	
removal	214
replacement	216
barrel and fore end	
removal	210–214
replacement	216–217
breech bolt and link	
removal	214–215
replacement	215
butt stock	
removal	214
replacement	216
carrier	
removal	214
replacement	215–216
trigger guard	
removal	214
replacement	216
operation (See Operation)	
Sear, Stevens shotgun, M620A, etc.	68
Shell	
description	7
Migratory Game Act, provision of for	7
removal from magazine and chamber	231
use of the word	3
Shell cut-off, Winchester, M12	39
Shell stop(s)	
Ithaca, M37	92
Remington, M11 and Sportsman	166
Remington, M31	142
Savage, M720	198

TM 9-285

SHOTGUNS, ALL TYPES

S—Cont'd   Page No.

Shell stop (s)—(Cont'd)
  Stevens, M620, etc. 62
  Winchester, M97 16
Slide
  Ithaca, M37 90
  Remington, M31 142
Slide and sliding breech, Stevens, M620, etc. 65–66
Slide handle
  Ithaca, M37 90
  nomenclature and parts 4
Slide lock
  malfunction, effect of 9
  Stevens, M620A, etc. 68
Slide stop, description, Ithaca, M37 92–93
Sliding breech description, Stevens, M620A, etc. 65
"Solid frame" defined 5
Sporting guns, use of 7
Sporting skeet and trap gun, classification of 6
Sportsman model (*See* Remington shotgun, M11 and Sportsman)
Standard nomenclature, use of 3
  ammunition 239
Stevens shotgun, M620A, etc.
  cleaning and lubrication 86–87
  data 69
  description 59–62
    general 62–68
    field inspection 84–86
      assembled parts, faulty 85–86
      general 86
      operation of gun 84–85
  functioning 73–75
  groups
    bayonet attachment, barrel, magazine, and operating handle
      removal 75
      replacement 83–84
    lifter
      removal and replacement 81
    slide and sliding breech
      removal 75–80
      replacement 83
    stock
      removal 80
      replacement 82–83

Page No.

trigger plate
  removal 80–81
  replacement 81–82
operation 69–73
  loading and unloading:
    gun 71
    loading the chamber only ... 73
    magazine 71
Stock
  Ithaca, M37 90
  Remington, M10 114
  Remington, M11 and Sportsman . 163
  Remington, M31 140
  Savage, M720 195
  Stevens, M620A, etc. 62
  Winchester, M12 39
  Winchester, M97 13
Stoppages and immediate action 224–231
  general discussion of 224
  malfunctions, cause and correction 226–231
  shell, removal from chamber and magazine 231
  stoppages and immediate action 224–226
Storage of ammunition 245
  (*See also under* Maintenance)

T

"Take-down" defined 4–5
Trigger, description, Stevens, M620A, etc. 68
Trigger guard group
  Savage, M720 200
  Winchester, M12 39
Trigger plate
  Ithaca, M37 90
  Remington, M31 142
  Remington, M11 and Sportsman . 167
  Stevens, M620A, etc. 66
  Winchester, M97 13
Trigger pull, testing 7–8
Testing assembled gun
  Ithaca, M37 108–109
  Remington, M10 133–134
  Remington, M11 and Sportsman . 189
  Remington, M31 157
  Savage, M720 219–220
  Stevens, M620A, etc. 84–85
  trigger pull 7–8
  Winchester, M12 55–56
  Winchester, M97 32–33

INDEX

T—Cont'd

	Page No.
Trigger safety	
Ithaca, M37	93
Remington, M31	143
Trigger safety lock, Winchester, M12	41
Types of guns (*See* Classification)	

U

Unloading (*See under* Operation)

W

	Page No.
Winchester shotgun, M12	
cleaning and lubrication	58
data	41
description	38–39
general	39–41
field inspection	55–58
assembled parts, faulty	56–57
general	57–58
operation of gun	56
functioning	46–47
groups	
barrel, magazine and action slide group (take-down gun)	
removal	47
replacement	55
bayonet attachment, barrel, magazine, and action slide groups (solid-frame gun)	
removal	47
replacement	55
breech bolt	
removal and replacement	54

	Page No.
trigger guard	
removal	47
replacement	54–55
Winchester shotgun, M97	
cleaning and lubrication	34–37
data	16–17
description	10–13
general	13–16
field inspection	32–33
assembled parts, faulty	33–34
general	34
operation of gun	32–34
functioning	20–23
groups	
barrel replacement	30–32
bayonet attachment (solid-frame gun)	
removal	24–25
replacement	32
breech bolt	
removal	28
replacement	30
carrier	
removal	25–28
replacement	30
magazine and action slide (solid-frame gun)	
removal	25
replacement	30–32
trigger guard bow	
removal	28
replacement	28–30
hammer gun	4
operation (*See* Operation)	

[
A.G. 062.11 (10-7-42)
O.O. 461/18885 O.O. (10-22-42)
TT GRA WAO 12 Dec. 071945Z
]

By order of the Secretary of War:

                        G. C. MARSHALL,
                                  Chief of Staff.

Official:

    J. A. ULIO,
        Major General,
        The Adjutant General.

Distribution: R 9(2); IBn 9(1); IC 9(4), 11(30), 19(8).
        (For explanation of symbols, see FM 21-6)

www.ingramcontent.com/pod-product-compliance
Lightning Source LLC
Chambersburg PA
CBHW031102080526
44587CB00011B/789